Co

An
Egypt

David Pickering

GW00367146

First published in 2007 by
Collins, an imprint of
HarperCollins Publishers
77-85 Fulham Palace Road
London, W6 8JB

www.collins.co.uk

Collins Gem is a registered trademark of HarperCollins Publishers Limited.

Reprint 10 9 8 7 6 5 4 3

Text © 2007 Collins

All photographic images supplied by Corbis
Illustrations on pages 42, 46-56, 65-66, 68-71, 168-176, 178,
214, 228 and 233 © 2007 Cara Wilson

A catalogue record for this book is available from the British Library

Created by: SP Creative Design
Editor: Heather Thomas
Designer: Rolando Ugolini

ISBN-10: 0-00-723163-6
ISBN-13: 978-0-00-723163-8

Printed in China

CONTENTS

INTRODUCTION

The civilization of the ancient Egyptians has fascinated the popular imagination since serious interest was first paid to the archaeological remains of Egypt during the nineteenth century.

CRACKING THE HIEROGLYPHIC CODE

For many centuries, knowledge of Egyptian civilization was incomplete, depending largely upon accounts left by Greek historians in classical times, and mystery surrounded the pyramids and other relics of Egypt's distant past. However, late in the nineteenth century, after much study, the curious hieroglyphics that the ancient Egyptians left behind them in tombs and temples were decoded, opening the door to a greater understanding of a society that was always known to have been rich in cultural and scientific achievements.

Daily life in ancient Egypt

The details of daily life of the people living on the banks of the Nile some 3,000 years ago have become accessible to us through the work of generations of archaeologists and other Egyptologists. Long-forgotten pharaohs, together with their mummified remains, have re-entered popular folklore and have become the inspiration for numerous plays, novels and films as well as historical studies. Egyptian hieroglyphics and artistic styles have become widely familiar, and in the 1960s and 1970s millions of people round the world flocked to see the fabulous treasures found in the tomb of Tutankhamun when they were exhibited on tour.

Ancient Egyptian culture

The amazing state of preservation of many objects (including mummified bodies) dating back thousands of years has brought modern generations into direct contact with the ancient Egyptians, making their world seem much more real and immediate to us than many other societies of equal antiquity.

As a result, the cultural influence of the ancient Egyptians continues to be felt even today, while Egypt itself attracts thousands of tourists every year, all eager to see the monuments, tombs and other sites that still bear witness to one of the world's earliest and most fascinating great civilizations.

Dating

It should be noted that experts and archaeologists continue to debate the exact dates of events in ancient Egyptian history. The dates given in this book reflect current opinion about when things happened, but other books may offer slightly different dates.

PART ONE

The land of the pharaohs

The emergence of civilization in Egypt depended heavily upon the region's geographical features. Most important of these was the Nile, the longest river in the world, on the banks of which the first Egyptian settlements and cities evolved and developed.

The River Nile

Life in ancient Egypt relied upon the River Nile and its annual flood, which watered the land along its banks and made it suitable for farming.

THE VALLEY OF THE NILE

Little rain falls in the north-west corner of Africa where Egypt is located. Without the River Nile, Egypt would be little more than an uninhabitable hot sandy desert, and there would never have been an ancient Egyptian civilization, which sprang up along its banks.

Many thousands of years ago, Egypt was mainly swampland, but as the climate grew drier the Nile retreated, leaving a strip of fertile soil on both banks. This soil was refreshed each year by the floodwaters that covered the fields with rich black mud in which plants grew readily. The early settlers learned to farm this land and to construct canals and dykes to regulate the flow of water. Towns sprang up along the length of the Nile, the river providing the people with drinking water as well as water for crops.

THE BLACK LAND

The Egyptians called the fertile land on which they built their towns Kemet, meaning 'black land'. The rest of the country was known as the Red Land. The deserts of the Red Land were very important because they protected the early Egyptians from attacks by neighbouring civilizations, as few armies were willing to cross such desolate waterless wastes.

Sacred river

The ancient Egyptians understood that their lives depended heavily upon the Nile. In their paintings, heaven was often depicted as fertile land surrounded by water.

The waterfalls and rapids of the Nile south of Egypt also hindered invasion from lower down the river. The Egyptians themselves transported their soldiers up or down the Nile to fight off any invaders. The Nile was also vital for trade, providing access to the waters of the Mediterranean Sea in the north and to central Africa in the south.

EGYPT AND THE WORLD

The geographical location of Egypt at the point where Africa meets the Middle East was to prove vital to its development into a great civilization. The natural resources of the region, which included minerals and building stone from the desert as well as crops from the fertile fields beside the Nile, could be transported either by river or across the desert to be traded with neighbouring peoples.

Access to the Mediterranean and to the Red Sea enabled Egyptian merchants sailing out of the Nile to take their goods anywhere in the known world.

EMPIRES OF THE PHARAOHS

With economic development came great wealth and political influence. The physical size of Egypt's empire varied from age to age, sometimes extended by military conquest – at its greatest extent, it included large parts of northern Africa and Palestine among other neighbouring regions. The relatively close geographical links with the centres of Greek and Roman civilization were particularly important in the later history of ancient Egypt.

Centres of civilization

The most important centres of ancient Egyptian civilization were the traditional capitals of Memphis in the north (including the pyramids at Giza, near modern Cairo) and Thebes in the south (with such associated sites as Karnak, Luxor and the Valley of the Kings). The last capital of ancient Egypt, Alexandria (on the Nile delta in the far north of the country) was not built until the fourth century BC.

Aegyptus to Egypt

The word Egypt is derived from the Greek name for the country: *Aegyptus*. In Greek mythology, Aegyptus was a descendant of the heifer maiden *Io* and was the country's first ruler.

Ancient Egypt

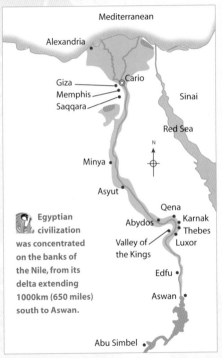

Mediterranean

Alexandria

Cairo

Giza
Memphis
Saqqara

Sinai

Red Sea

N

Minya

Asyut

Qena

Abydos — Karnak
Thebes
Valley of — Luxor
the Kings

Egyptian
civilization
was concentrated
on the banks of
the Nile, from its
delta extending
1000km (650 miles)
south to Aswan.

Edfu

Aswan

Abu Simbel

PART TWO

The history of ancient Egypt

The civilization of ancient Egypt lasted some 3000 years, longer than any other civilization in world history. At its peak it was remarkable for its cultural, religious and economic achievements. Historians traditionally subdivide this era into a dozen or so periods covering 30 ruling dynasties, beginning around 3000 BC and ending with the Roman invasion in the first century AD.

Egyptian life
A man and his family hunting wild birds on the River Nile, which was always at the heart of ancient Egyptian civilization.

PREDYNASTIC PERIOD

c. 8000 BC The drying-out of the Sahara caused by a change in the climate, and possibly over-grazing, forces the population to concentrate in the Nile Valley. Egyptian farmers become the first to herd domesticated cattle and other livestock.

c. 6000 BC The first simple single-sailed ships are built. Stone tools are replaced by tools made of metal. Trades include tanning and basket-weaving.

c. 5000 BC Two broadly similar civilizations develop, one in the south of the country (Upper Egypt) and one in the north (Lower Egypt).

c. 4000 BC The first large stone buildings are constructed. Advances are also made in the fields of alchemy, cosmetics, music, medicine and pottery. The building of underground tombs containing grave goods anticipates later burial practices.

c. 3500 BC The Egyptians invent senet, the world's oldest known board game.

The first people came to Egypt from central Africa some time between 100,000 and 50,000 BC. Archaeological finds suggest that Egypt's original hunter-gatherers learned to grow grain beside the Nile by 10,000 BC. It was on this prosperous agricultural economy that the country's later greatness was to depend.

FLINT TOOLS

The first Egyptians belonged to wandering tribes who relied upon the use of flint tools. As the centuries passed they gradually settled in the Nile Valley, living by hunting animals and fishing.

CRAFTS AND TRADE

Around this time the Egyptians also developed skills in using stone, copper, clay, wood and leather and established thriving craft industries. The building of simple papyrus boats promoted links between communities and enabled traders to travel up and down the Nile. Villages and towns sprang up along the length of the river, from the delta in the north to what is now Aswan in the south, each with their own tribal chief, gods, burial customs and artistic styles.

UPPER AND LOWER EGYPT

Little is known about the history of these early kingdoms, though several rulers are known by name, among them Crocodile, Scorpion I, Scorpion II, Iryhor and Ka. The eventual merging of these two kingdoms, which shared a common language and system of writing, was to result in the first true nation and the beginning of what is generally considered ancient Egyptian civilization.

EARLY DYNASTIC PERIOD

c. 3200 BC The Egyptians perfect their hieroglyphic writing system.

c. 3100 BC The union of Upper and Lower Egypt creates a single state extending from the Mediterranean in the north to Aswan in the south.

c. 3000 BC Papyrus is first used for writing.

c. 2750 BC Fortunes decline following civil war between rival supporters of the gods Horus and Set.

Most of today's scholars date the beginnings of Egyptian civilization to the emergence of a unified Egypt under the rule of a single pharaoh in the fourth millennium BC. By this time the foundations for future greatness had been laid, including extensive trading links, well-organized government and advances in the arts and sciences.

GOVERNING EGYPT

The first pharaohs ruled from a new capital, at Memphis in the north. Notable events of this period included the development of trade with Palestine and Nubia and advances in craftsmanship, notably in the handling of hard stone.

FIRST OF THE PHARAOHS

The first ruler (or pharaoh) of united Egypt is popularly identified as King Menes, although historically King Narmer may have a better claim to this honour.

BURIAL CUSTOMS

The first pharaohs and other important people were buried in cemeteries at Abydos, and later at Saqqara, often in mastabas – flat-roofed rectangular mudbrick tombs.

 The Narmer Palette which depicts the first pharaoh of a united Egypt killing a defeated enemy.

OLD KINGDOM

c. 2686 BC Egypt enters a golden age with the start of the period known as the Old Kingdom.

c. 2650 BC Egyptians perfect the mummification process and the first of the pyramids of Egypt is constructed.

c. 2600 BC The Old Kingdom reaches its peak under the rule of Dynasty IV.

c. 2589 BC Khufu, builder of the Great Pyramid at Giza, becomes pharaoh.

c. 2500 BC Khafra orders the construction of the Sphinx.

c. 2400 BC The Egyptians develop an astronomical calendar.

c. 2183 BC The Old Kingdom ends and Egypt breaks up into separate parts.

This was the first of three great eras in the history of ancient Egypt, witnessing spectacular cultural achievements and the extension of Egyptian influence through the ancient world. Under the rule of Dynasties III to VI, Egypt became the greatest civilization the world had yet seen. The power of the pharaohs was never stronger than it was during this period, and the godlike status of Egypt's kings was affirmed in the building of the pyramids and other monuments.

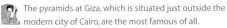
The pyramids at Giza, which is situated just outside the modern city of Cairo, are the most famous of all.

THE AGE OF THE PYRAMIDS

The Old Kingdom witnessed the building of Egypt's most famous pyramids. The first was the Step Pyramid at Saqqara, which was designed for the pharaoh Djoser by his vizier Imhotep.

Pyramid-building continued under many of the later pharaohs, notably Sneferu, builder of three pyramids, and Khufu and Khafra. Vast work forces were assembled to construct these massive monuments, which might take decades to complete. The pyramids at Giza remain the only survivors of the Seven Wonders of the World.

THE LAST OLD KINGDOM PHARAOH

Four hundred and fifty years of Egyptian stability and prosperity finally came to an end with the reign of the pharaoh Pepi II. Pepi II came to the throne of ancient Egypt at the tender age of six and he is said to have been over 100 years old when he died, making his reign the longest in Egyptian history.

Undermining the pharaoh

However, as Pepi II grew old and feeble, the Egyptian priesthood and nobility were increasingly empowered and they took advantage of the king's advancing years and weakness, and the office of the pharaoh was fatally undermined.

FIRST INTERMEDIATE PERIOD

c. 2183 BC Egypt descends into chaos, famine and civil war.
c. 2046 BC Civil war ends when the pharaoh of Upper
Egypt defeats the pharaoh of Lower Egypt and assumes
power as Mentuhotep II.

The ambitious building projects of the Old Kingdom,
coupled with civil war and a failure of the flooding of
the River Nile due to a cooling in the climate, led to a
period of famine and strife which lasted through the
reigns of Dynasties VII to XI. It was during this time that
most of the pyramids and tombs of earlier pharaohs
were plundered by tomb-robbers.

MANY GODS

It was during the turbulence of the First Intermediate
Period that the number of gods who were worshipped
by the ancient Egyptians increased, in the belief that
different gods ruled over different aspects of a person's
life and thus each one had a different character and
role to play. For more detailed information about the
various gods – there were over 2000 of them in total –
and also the ancient Egyptians' religious beliefs, turn
to page 42.

FAMINE

For around 60 years during the First Intermediate Period, the prosperity and cultural achievements of the Old Kingdom became little more than a memory as rival capitals were established at Herakleopolis Magna in Lower Egypt and at Thebes in Upper Egypt.

For ordinary Egyptians, it was a struggle to survive during this time of hardship and famine – indeed, an inscription in a tomb dating from this period describes how hunger even forced some people in Upper Egypt to eat their own children.

Lamentations

Conditions in ancient Egypt during this period are hinted at in an old text which is called 'Lamentations': 'The land is full of rioters. When the ploughman goes to work he takes a shield with him. The Inundation is disregarded. Agriculture is at a standstill. The cattle roam wild. Everywhere the crops rot. Men lack clothing, spices, and oil. Everything is filthy: there is no such thing as clean linen these days. The dead are thrown into the river. People abandon the cities and live in tents. Buildings are set on fire, though the palace still remains. But pharaoh has been kidnapped by a mob. The poor have become rich… Luxury is rampant, but the ladies of the nobility exclaim: "If only we had something to eat!"'

MIDDLE KINGDOM

c. 2046 BC The period known as the Middle Kingdom opens with the reunification of Upper and Lower Egypt.

c. 2000 BC Egyptian physicians write the world's first medical textbook.

c. 1800 BC The Egyptians reconquer northern Nubia and send armies into Palestine.

c. 1759 BC The Middle Kingdom ends with the death of the female pharaoh Sobekneferu.

Egypt entered its second golden age with the dawn of the Middle Kingdom, which spanned Dynasties XII to XIV. During this period in its history, threats from neighbouring countries were silenced and Egypt became the most powerful country in the region.

TRADE LINKS

During the Middle Kingdom trade links were greatly improved with Egypt's neighbouring countries. Profitable mining operations were developed in the Sinai desert and Egypt became increasingly wealthy. This wealth provided the funds that were needed for many ambitious building projects, which included a fine temple at Abydos (now lost).

A GOLDEN AGE

Overseeing this golden age in ancient Egypt were the pharaohs of Dynasty XII. Outstanding among them were Amenemhet I, who brought local governors back under pharaonic control; Senusret I, who recaptured lost territory and made Egypt stable and prosperous; the great warrior-pharaoh Senusret III, who strengthened Egyptian defences in Nubia; and Amenemhet III, who consolidated the power of the royal court.

Invention of writing

Among the most significant advances in this period was the introduction of a standardized writing system. Scholars compiled the first major body of literary texts during the Middle Kingdom (although some were attributed to the authors of the Old Kingdom).

Foreign settlers

Egyptian prosperity during the Middle Kingdom attracted increasing numbers of settlers from the neighbouring countries. They contributed to the national wealth through the payment of taxes and tariffs, but these new communities of foreigners stayed loyal to their own laws and leaders. As the settlements grew in size and power, their leaders assumed the status of kings, challenging the authority of the pharaoh. As a result, Egypt descended into a period of strife and disorder, and pharaonic power was undermined.

SECOND INTERMEDIATE PERIOD

c. 1759 BC Egypt enters another troubled period under a succession of weak pharaohs.
c. 1674 BC Much of Egypt is invaded by Asiatic tribes called the Hyksos (a word meaning 'Princes of Foreign Lands').

Ancient Egypt slipped into confusion once more at the close of the Middle Kingdom. For 200 years the country was plagued by internal division as well as invasion by foreign armies.

OCCUPATION BY THE HYKSOS

The Hyksos overran the northern part of Egypt and established their capital at Avaris. Their leader Salitis became the first pharaoh of Dynasty XV. Later Hyksos pharaohs held power as Dynasty XVI, although the Egyptian pharaohs of Thebes resisted and claimed authority as Dynasty XVII. Eventually, the Thebans recaptured Avaris and the Hyksos were expelled.

Chariots and bows

However, the influence of the Hyksos was not entirely negative. They are believed to have introduced horse-drawn chariots and more powerful bows to Egypt.

NEW KINGDOM

c. 1548 BC Egypt enters a new age of prosperity and cultural achievement called the New Kingdom.

c. 1500 BC The world's first glass-making takes place in Egypt.

c. 1360 BC Egypt's empire reaches a peak under the pharaoh Amenhotep III.

c. 1300 BC The shaduf device to irrigate fields with water from the Nile is introduced.

c. 1160 BC Egyptian scholars draw up the world's first known maps.

The New Kingdom witnessed the third and last great age in the civilization of ancient Egypt, lasting around 500 years. Under the rule of Dynasties XVIII to XX, Egypt enjoyed success both on the battlefield and in the marketplace and had a unique impact upon the early development of the arts and sciences.

EGYPT'S EMPIRE

The reconquest by Thutmose I of Nubia, Palestine and Syria led to the creation of an Egyptian empire. Thutmose III, remembered as 'the Napoleon of Egypt', extended Egypt's frontiers and added new territories

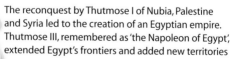

 The painted bust of Nefertiti is one of the great and most easily recognizable treasures of the New Kingdom.

in Asia. Other celebrated pharaohs of the New Kingdom included the female pharaoh Hatshepsut, who promoted trade with Egypt's neighbours.

RELIGION

Amenhotep II's son Akhenaten proved to be the most controversial of all the pharaohs. His reign witnessed the replacement of Egypt's old religion with the new worship of the sun-god Aten. These revolutionary if short-lived changes were accompanied by a new naturalistic style in art.

ART TREASURES

Many of ancient Egypt's greatest art treasures date back to the New Kingdom, among them the celebrated painted bust of Akhenaten's queen, Nefertiti, and the gold death mask of Tutankhamun.

Royal tombs

Most of the pharaohs of the New Kingdom were buried not in pyramids but in the Valley of the Kings. Many of their tombs were plundered by tomb-robbers in early times, but archaeologists have unearthed many significant finds in the area over the last 100 years, most notably the tomb of the boy-pharaoh Tutankhamun.

THIRD INTERMEDIATE PERIOD

c. 1070 BC–664 BC As the pharaohs decline in power, the priests become more powerful.

c. 818 BC Rival pharaohs rule in Lower Egypt and in Middle and Upper Egypt.

c. 732 BC The Nubians march north and defeat the combined armies of various native Egyptian rulers.

c. 700 BC Egypt is engaged in warfare with Assyria.

671 BC Egypt is invaded by the Assyrians.

The 400 years following the New Kingdom (known as the Third Intermediate Period) saw a reduction in the power of the pharaohs. Although there were periods of stability, this era in ancient Egyptian history also witnessed repeated outbreaks of civil conflict and, ultimately, invasion by foreign armies.

RIVAL DYNASTIES

The Third Intermediate Period saw Egypt ruled by Dynasties XXI to XXIV, though the exact order of rulers is often confused, with rivals ruling Upper and Lower Egypt respectively. Initially, the pharaohs of Lower Egypt struggled against the hereditary priest-kings of Upper Egypt, based at Thebes. Egypt's international standing plummeted with the Thebans failing to develop a foreign policy and the pharaohs proving too weak to act. The situation improved around 945 BC under Shosheng I who ushered in a period of stability.

FOREIGN INVASIONS

Disorder within Egypt encouraged the Nubians to the south to make repeated attacks and at various times Thebes and other areas came under Nubian control. Ultimately the Nubian pharaohs of Thebes sent their armies north and defeated the combined forces of native Egyptian rulers until all of Egypt was conquered.

REUNIFICATION

The Nubian rulers of Dynasty XXV spent much of their time fending off attack by Assyria, which had emerged as the chief rival to an Egypt in decline. Assyrian invasions resulted in the capture of both Thebes and Memphis and in the expulsion of the Nubian pharaohs. The country was thus reunified once more under Psamtik I, founder of Dynasty XXVI.

OLD-STYLE ART

Though turbulent, this lengthy period did leave behind sculptures and paintings, which typically harked back to the styles of the Old and Middle Kingdoms.

LATE PERIOD

664 BC Peace is restored and Egypt enters a new era of prosperity and cultural achievement.
525 BC Egypt is invaded by the Persians.
332 BC The Egyptians welcome liberation from Persian control when Alexander the Great of Macedon arrives.

After many years of instability Egypt enjoyed a late flowering under Dynasties XXVI to XXXI, although it was dominated by Persia for much of this time. Egyptian influence beyond its borders was, however, a shadow of the greatness it had achieved in earlier eras.

STRONG ECONOMY

Egypt was never again to return to the heights that had been reached under the New Kingdom, but this period witnessed a resurgence in native Egyptian culture and an improvement in national fortunes based on good social organisation and a strong economy.

PERSIAN PHARAOHS

The Persian king Cambyses assumed the title of pharaoh and became the founder of Dynasty XXVII.

Egypt remained a province of Persia until 404 BC, when a native Egyptian pharaoh returned to the throne. Egyptian rulers held power as Dynasties XXVIII to XXX, but a second Persian invasion in 343 BC toppled Nectanebo II, who is considered the last native Egyptian pharaoh.

EGYPTIAN RULE

From 332 BC the fortunes of ancient Egypt would be heavily influenced by pressure from the Greeks and, later, their successors, the Romans. It was not until the nineteenth century AD that another native Egyptian was to rule the country.

After the Persian army of Artaxerxes III defeated the Egyptians under Nectanebo II, Egypt became part of the Persian Achaemenid empire.

PTOLEMAIC PERIOD

332 BC Alexander the Great is proclaimed master of Egypt.
323 BC Alexander the Great dies and his general Ptolemy I Soter becomes pharaoh.
47 BC With the support of Julius Caesar, Cleopatra VII assumes the throne of Egypt.
30 BC The death of Cleopatra VII ends the Ptolemaic period.

Egypt was ruled by pharaohs of Greek descent for 300 years. Although it was a source of wealth and prestige to the Greeks and, later, the Romans, ancient Egypt never again achieved true independence.

EGYPT UNDER THE GREEKS

Most Egyptians, weary of oppression by the Persians, welcomed Alexander the Great's arrival. He showed due respect to the gods of Egypt, but then appointed Greeks to all the senior posts in the country, and Greek influence soon spread throughout Egypt.

A new capital city, which was named Alexandria after the great man himself, was founded on the coast of the

Mediterranean Sea in the north. Egypt was to remain part of the Greek world for hundreds of years, its riches funding Greek military campaigns.

THE PTOLEMIES

Ptolemy I Soter was the first pharaoh of a new dynasty that was to last 300 years. All of Macedonian descent, the Ptolemies adapted to Egyptian customs, marrying within the family to keep their bloodline pure and building new temples to Egyptian gods.

The Ptolemies lived in lavish surroundings at court, and some pharaohs proved themselves capable defenders of Egypt's interests, leading Egyptian armies into battle or dedicating themselves to supporting and patronizing science, literature and the arts.

THE END OF THE PHARAOHS

Weakened by repeated intermarriage, many of the later Ptolemies proved to be deranged, corrupt or just plain incompetent. Ptolemy X, for instance, was lynched by a mob in Alexandria after murdering his mother. Egypt eventually became a province of Rome, upon which it had become increasingly reliant over the centuries. The last of the Ptolemies, Cleopatra VII, was not only a key figure of her time but won enduring fame as the most romantic and ultimately tragic ruler of ancient Egypt.

ROMAN PERIOD

30 BC Egypt is conquered by the Romans under Octavian.
c. AD 100 Christianity is introduced to Egypt.
AD 395 Egypt becomes part of the Byzantine (Eastern Roman) Empire.

The history of ancient Egypt is usually considered to have ended with Roman invasion, after which the country became a province under the control of the emperors in Rome. Several features of Egyptian culture, however, continued to flourish for many years, though eventually the Roman and, later, the Arab influences became dominant.

ROMAN EGYPT

As a province of the Roman Empire, Egypt was ruled by a governor appointed by the emperor. Rome thus secured the valuable Egyptian grain supply. Romans replaced Greeks in the senior government posts but most of the country's business continued to be done in Greek and the Greek influence remained dominant. No attempt was made by the Roman rulers to suppress the Egyptian religion and traditional customs, although the cult of the emperor was stressed and many new buildings were constructed in the Greco-Roman style.

THE INTRODUCTION OF CHRISTIANITY

With the introduction of Christianity the old religion and gods of Egypt started to retreat. The third century AD witnessed civilian and military revolts as well as repeated persecution of Christians, whose numbers continued to grow. Christianity became the religion of the Roman Empire in the fourth century, but there is evidence that the old religion continued in isolated pockets into the fifth century.

A NEW EGYPT

Over 900 years of Greco-Roman rule finally ended with the conquest of Egypt by the Persians in 616 and by the Arabs in 639.

Lost in the sands

The replacement of the pagan religion of the ancient Egyptians by Christianity led to the disappearance of the old Egyptian priesthood. With them went the ability to understand hieroglyphics, which remained a secret for some 1500 years. Their temples, meanwhile, vanished beneath the desert sands only to be uncovered once more by archaeologists from the nineteenth century onwards.

PART THREE

The gods of ancient Egypt

Religion was very important to the ancient Egyptians. The pharaohs were considered gods and much effort was expended on erecting magnificent temples and monuments to the divine beings on whose favour the country's fortunes were believed to depend.

The scarab

Treasures found by archaeologists have included numerous images of the gods as well as objects depicting sacred symbols or animals believed to have magical luck-giving or protective powers.

RELIGIOUS BELIEFS

The ancient Egyptians believed that the otherwise mysterious workings of nature, including the flooding of the Nile, were controlled by a gallery of gods to whom it was essential to show due respect. They worshipped over 2000 different gods, each of whom had a different character and role. The gods were

The bull god

Examples of revered local gods included the bull god Apis, who was worshipped in the capital city of Memphis. Apis was depicted as a bull bearing a sun between its horns.

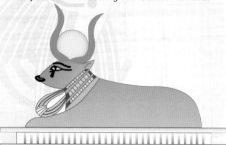

typically depicted as being half-human and half-animal (or sometimes as all-animal), often having the head of an animal mounted on a human body. In some cases the animals with which the gods were identified varied from temple to temple, as did their names.

LOCAL GODS

Each city or area (*nome*) of ancient Egypt had its own god. Many gods were known only locally and in some cases were associated with a particular hill or a tree. These minor deities were honoured at small shrines made of mud bricks or matting, in which was housed a clay statue representing the relevant god. The worshippers prayed at these shrines and made their requests for divine favours.

More important gods were worshipped in large stone temples housing statues of the gods fashioned in gold or other precious materials by skilled craftsmen.

Magical names

The ancient Egyptians gave their gods their own unique appearance, character and as many as five different names, which were considered to have powerful magical properties. It was widely believed that by simply writing the name of an enemy on something, and then breaking it, the person in question would suffer some misfortune or even die.

THE KING OF THE GODS

A relatively small number of gods, under the lordship of a single creator-god, were revered throughout the country through the entire history of ancient Egyptian civilization. Most important of all these was the sun

Shaven-headed priests of Amun, depicted here in a wall painting found in a tomb dating back to the reign of Ramesses III in the twelfth century BC.

The rebel god

Under the pharaoh Akhenaten, worship of Amun-Re and the other old gods was replaced by worship of a single creator-god called Aten. Aten was symbolized by a disc with rays ending in human hands holding the sign of life out to the royal family. The old gods, however, were later restored under Tutankhamun.

god Re (or Ra), the king of the gods and traditionally the first ruler of Egypt, from whom the pharaohs claimed descent as gods on earth. Re was revered as the lord of all creation, who also ensured the fertility of the soil and escorted dead kings into the underworld. It was believed that each evening the sun god was swallowed by the sky goddess Nut, only to be reborn the following morning after travelling through the underworld and thus bringing life back to the world after each night.

AMUN-RE

Little was known about the character of this supreme god, who was later worshipped as Amun-Re (Amun meaning 'hidden'). He was depicted in several different guises, which included Khepri (a scarab beetle busily rolling the sun over the horizon at dawn) and the falcon-headed Re-Harakhty.

GODS AND GODDESSES

Under the lordship of Re or Amun-Re was a host of other lesser gods and goddesses. These gods were worshipped at all levels of society and their images survive in temple ruins and in ancient inscriptions.

NUT

The daughter of Shu – the god of the air – Nut was the sky goddess who married her brother Geb, the god of the earth, and gave birth to Isis and Osiris. She is usually depicted with her body decorated with stars, arched over the earth with her feet planted in the east and her hands in the west.

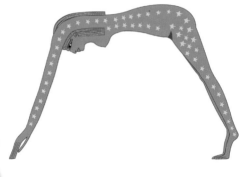

OSIRIS

The son of Geb and Nut, Osiris was murdered by his jealous brother Seth, but restored to life by Isis as ruler of the underworld, where the dead reside. He thus symbolized both the afterlife and the rebirth of the land after the Nile flood. He is usually depicted wearing a crown of reeds and ostrich feathers and holding flail sceptres as a mark of kingship.

ISIS

The wife and sister of Osiris, Isis brought her husband back to life after his murder. The mother of Horus, she was worshipped as a mother goddess and as the protector of children. She is often depicted with the wings of a kite (which enabled her to fly in search of her murdered husband's mutilated body).

HORUS

The son of Isis and Osiris, Horus was a falcon-headed sky god who successfully fought Seth for the throne of Egypt. The pharaohs claimed to be his reincarnation and were often depicted with the falcon head of Horus. Horus himself was sometimes depicted as a child. He was believed to protect households from natural threats, such as crocodiles, snakes and scorpions.

SETH

The god Seth seized the throne of Egypt by murdering his brother Osiris and then scattering his body along the Nile. However, he was then defeated by Horus. He is the god of Chaos and is usually depicted as an aardvark-like creature. Seth was also associated with the male hippopotamus, and an important ancient Egyptian ritual involved the reigning pharaoh killing a hippopotamus with a harpoon.

BAST

The daughter of Re, Bast (or Bastet) was depicted as a black cat, and she was worshipped as the goddess of cats, musicians and dancers. Cats were sacred in ancient Egypt and when dead were buried in cat-shaped coffins in cemeteries that were specially for cats. Bast symbolized the power of the sun to ripen crops.

ANUBIS

The god of the dead, Anubis was closely associated with the mummification process. He was depicted with the head (and also sometimes the body) of a jackal, probably because jackals were commonly found near cemeteries and were believed to protect the dead. Priests wore Anubis masks when preparing mummies.

HATHOR

The goddess of pleasure, music, dancing, women and love, Hathor had the ears of a cow, which was viewed as a gentle, loving creature. She was the ideal of beauty and her great temple at Dendera was decorated with huge heads depicting her face.

THOTH

The moon god Thoth was variously depicted with the head of an ibis (a bird whose curved beak resembled the crescent moon) or as a baboon. The god of both wisdom and healing, he was supposed to have brought hieroglyphics, medicine and mathematics to Egypt and was the patron of scribes.

SOBEK

The god of the crocodile-infested waters of the River Nile, Sobek was always depicted with the head of a crocodile. Sacred crocodiles were kept in pools beside his temples, and dead crocodiles were sometimes mummified. Worship of Sobek was particularly intense at the city of Arsinoe, which was eventually renamed Crocodilopolis by the Greeks.

KHNUM

The ram-headed god of the River Nile's rapids and waterfalls was Khnum. It was said to be at his command that the god Hapy caused the annual flooding of the Nile. The Egyptians believed that Khnum decided how much silt was left on the land after each flood, and he was thus worshipped as a fertility god. He was also revered by Egypt's potters, who used the mud of the Nile to make their pots.

TEMPLES

The ancient Egyptians honoured Amun-Re and other important gods and goddesses by building huge temples dedicated to their worship. Many of these temples survive as magnificent ruins to this day.

TEMPLE COMPLEXES

Most temples shared the same basic design, which comprised a grand entrance (called a pylon), behind which was an open courtyard and a second smaller pylon. This opened onto a second courtyard, and then a hall filled with massive columns supporting a roof (a hypostyle hall). This might open into a smaller hall (a hall of offerings) before reaching the innermost sanctuary in which was kept the sacred statue of the god to whom the temple was dedicated. This statue was often made of gold, which was held sacred as the flesh of the sun god Re. The various halls and courtyards were adorned with further statues and paintings of the gods.

Close shaves

Visitors to the temples of ancient Egypt had to shave off their hair and eyebrows before entering, in order to preserve the sanctity of the buildings.

Temple complexes consisted not only of the central place of worship, but also the neighbouring buildings housing schools, libraries, store-rooms and living quarters for priests, craftsmen and other workmen.

Examples of impressive temples that are still standing in a more or less intact state include those at Karnak, Luxor, Edfu, Abydos and Dendera. One of the most

celebrated is the colossal hillside temple at Abu Simbel, built by Ramesses II to honour both Re and Ramesses himself. Guarded over by four vast statues of Ramesses, the temple was designed so that twice a year the sun's rays would penetrate an inner shrine.

 The temples not only proved faith in the gods, but also demonstrated the wealth and power of the pharaohs.

RITUALS

The religious ceremonies that were carried out in the temples of ancient Egypt were often highly elaborate and could involve the whole population of the area. At the major ceremonies, the pharaoh himself might preside over the proceedings, in which burning incense, holy water, music and offerings of food and other objects were common features.

Sed-festival

The most important of these ceremonies included the Sed-festival, a royal jubilee celebrated in the thirtieth year of a pharaoh's reign that emphasized the divine powers of kingship. During the ceremony the nobility renewed their pledges to the pharaoh and re-enacted his coronation. Other festivals marked the beginning of the New Year or were held in honour of various gods.

Lesser rituals

These rituals were observed on a daily basis under the supervision of the high priest of the temple. Each morning the priests ceremonially broke a clay seal on the door leading to the innermost sanctuary in which the god's statue was kept. They would then make a daily offering of food and drink to the god's sacred statue. The offerings themselves were carefully washed in sacred water to ensure their purity.

FEAST DAYS

Each of the gods had a special feast day, which was marked with processions, feasting and dancing. On these – and on other special days – the god's statue would be brought out of the temple and then paraded through the streets by standard-bearing priests.

On these occasions the ordinary Egyptian people, who would be granted a festival holiday, would have the opportunity to consult the god, although the statue itself was usually kept hidden from view in a golden casket or similar covering.

Humble offerings

Ordinary people were not usually allowed into the temples themselves as they were thought unworthy of such honour, but they would be permitted to leave offerings in the temple courtyard, including food for the priests.

Festival calendars

Detailed lists of festivals were inscribed on the walls of many of ancient Egypt's temples. These included instructions detailing the date on which a festival should be held, together with the name of the god to be honoured and information about the rituals to be performed.

HOUSEHOLD GODS

While the most important gods were worshipped in big temples, some of the lesser gods had a special place in the affections of Egyptian families and were worshipped in their homes. Every house had a shrine devoted to such worship, often a bricked-off part of the main front room. Here members of a family could pray for divine help or leave food for the spirits of their ancestors in the hope that they would help protect the living from any threat of harm. Expectant mothers would also be placed in the shrine when childbirth was imminent in the belief that various household gods would keep any evil spirits away.

FAMILY GUARDIANS

Among the most important of these family gods were Taweret, the goddess of childbirth who was commonly depicted in the guise of a pregnant hippopotamus, and the dwarf-god Bes, the guardian of the family and the protector of the newborn in particular, whose grotesque features were believed to keep evil at bay.

Many houses were also protected by inscriptions that were scratched on tablets of stone or other material in which the family called upon the assistance of Horus and other deities. Many examples of these tablets (called *stelae*) have been found.

PRIESTS

The priests of ancient Egypt occupied a privileged position in society, and indeed the power of some high priests rivalled that of the pharaoh himself. Each temple had a high priest who represented the pharaoh in important religious ceremonies when the latter was unable to attend. Priests in ancient Egypt did not preach to congregations or act as messengers of divine truth to the general population. Instead, it was their duty to retain the favour of the gods by pleasing them with their ritual observances. The high priest oversaw all the rituals and associated activities of the temple and had access to the huge wealth contributed to the temples by the rest of the population in the form of taxes. The office of high priest was usually passed down from father to son unless the pharaoh himself intervened and appointed someone else.

DIVISION BY RANK

Officials of the temples were treated with great respect and a priest might be referred to as 'God's servant' or 'Pure one'. The priesthood was divided into various ranks, denoting each man's level of seniority. The humblest priests were responsible for more mundane duties, such as sweeping the temple floor or looking after administrative records.

SHAVEN HEADS

Priests had to keep themselves very clean, and if they were dirty in any way they were considered unworthy to enter the gods' presence. To keep themselves pure, they bathed twice each day and twice each night in sacred pools. They also shaved their heads and often removed all other body hair as well.

The priests also had duties outside the temple itself. These included teaching the boys of wealthier families in the temple schools. When engaged in such activities they wore a woollen wig over their shaved head.

FAMILY LIFE

Priests were allowed to marry and enjoy normal family life. Most served as priests for just one month in three; the rest of the year they pursued careers in business, administration and other professions.

Women priests

Few women were admitted to the priesthood. However, some reached senior positions, such as high priestess of Amun in Thebes (the most important person in the land after the pharaoh himself).

SACRED SYMBOLS

The ancient Egyptians believed that certain symbols communicated divine power. They appear repeatedly in inscriptions, in temples and in tombs throughout Egypt as well as on jewellery, on coffins and on other artefacts now preserved in museums.

ANKH

The ankh symbolized life itself and it was often included in wall paintings and statues, although only gods and royalty (who alone wielded the power of life and death) were allowed to be depicted with it. The sun disc emblem of Aten, the supreme god who was introduced by pharaoh Akhenaten, included the ankh symbol.

WADJET

Otherwise known as the eye of Horus, the wadjet – or wedjat – had its roots in the legend of the battle that was fought by the gods Horus and Seth for the throne

of Egypt, during which Horus lost an eye. The eye was later magically restored and became a sacred symbol, which was believed to protect anything behind it.

IBIS

The ibis was considered to be sacred by ancient Egyptians as the symbol of Thoth, the god of wisdom and healing.

SCARAB BEETLE

The sun god Re, or Amun-Re, was sometimes depicted as a scarab beetle, rolling the sun through the sky in the same way that real beetles roll balls of dung along the ground. Lucky scarab amulets were believed to afford their wearer some protection. A variety of the emblem was the winged scarab, which was believed to have similar powers.

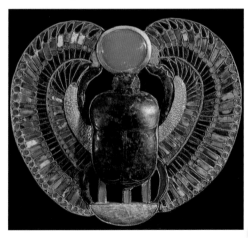

 Scarab amulets were often placed on the chest of mummified bodies to ensure their safety after death.

GIRDLE OF ISIS

The girdle of Isis (or tyet) was used to summon the aid of the goddess Isis. It is thought to depict a knot of cloth. The symbol is often found in funerary inscriptions, probably because Isis (the wife of Osiris) was believed to protect the spirits of the dead.

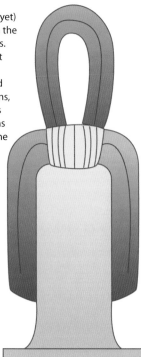

DJED

The djed (or pillar) was the symbol of the god Osiris and was revered as a symbol of stability, survival and also the afterlife. The origins of the djed, one of the oldest sacred symbols of the Egyptians, are obscure. Originally, it may well have represented a pole or leafless tree and been revered as an emblem of fertility.

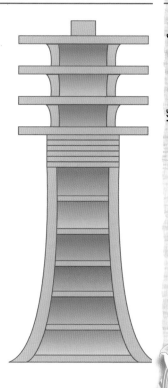

COBRA

The cobra snake symbol represented Lower Egypt and appeared in the crown of Egypt and in the jewellery of the pharaohs. Otherwise called the Uraeus, the cobra symbol was associated with the sun and was believed to have protective powers.

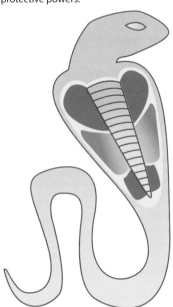

LOTUS

The lotus flower (common on the banks of the Nile) was the symbol of Upper Egypt and was consequently included in the architecture of important buildings and in wall paintings. Because the lotus (or sesen) closes each night and reopens at dawn, it was a symbol of creation and rebirth.

PART FOUR

The pharaohs

The kings of ancient Egypt were the pharaohs, who were worshipped as living gods. Of the hundreds of pharaohs who occupied the Egyptian throne, most are now forgotten. A few of them, however, achieved a fame that was to last thousands of years.

The lives of the pharaohs

The history of ancient Egypt is uniquely linked to the lives of the pharaohs, and many of the treasures kept in museums today were recovered from royal tombs.

THE ROLE OF THE PHARAOHS

The word 'pharaoh' originally meant 'great house' and referred to the magnificent palaces in which the kings of Egypt lived. The pharaoh was the most powerful person in the country and had the power of life and death over his subjects. Worshipped as living gods by ordinary people, the pharaohs were at the heart of ancient Egypt's political, religious and cultural life.

The title of pharaoh usually passed from a father to his eldest son by the most important of his wives, although very rarely it went to a female relative or to a trusted adviser. Thus most pharaohs belonged to dynasties of rulers all belonging to the same family.

REGALIA

The pharaohs wore crowns and other marks of office. The kings of Lower Egypt wore a bucket-shaped red

False beards

Because the gods were supposed to have beards, the pharaohs similarly attached false beards to their chins. Even Hatshepsut, one of the rare female pharaohs, wore one of these beards.

The pharaoh Senusret I, holding the ankh symbol representing power over life and death.

crown, while those of Upper Egypt wore a bottle-shaped white crown. When the two regions were unified, the ruler of the combined country wore a crown in which the red and white crowns were similarly merged.

Pharaohs of the New Kingdom, meanwhile, wore a blue crown resembling that of the sun god. Alternatively, pharaohs were depicted wearing a distinctive striped headdress called a nemes. Crowns and headdresses were sometimes decorated with either the cobra goddess of Lower Egypt or the vulture goddess of Upper Egypt, or both.

Other symbols of the pharaohs' power included the hand-held hook and flail – the emblems of the god Osiris which were identical to the implements used by peasants to separate grain from stalks.

THE LIFE OF A PHARAOH

Most pharaohs were trained for their future role as leader of ancient Egypt from their childhood. The long-established tradition of reigning pharaohs sharing their thrones with their sons helped to ensure the eventual smooth handover of rule from father to son when a pharaoh died.

ROYAL DUTIES

A pharaoh was the commander-in-chief of the army as well as being responsible for the administration of the country and for representing his people to the gods in the temples. When he was not engaged in warfare, the pharaoh might be busy dealing with economic matters, with his religious duties or with planning and overseeing the massive building projects of ancient Egypt, such as constructing temples and pyramids, which were considered to be necessary for retaining divine favour both in this life and the next.

Longest reign

The longest-lived pharaoh of all was Pepi II, who lived to be around 100 years old and ruled for some 94 years. His reign remains the longest in recorded history.

Attending ceremonies

The pharaoh also had to attend ceremonial occasions, which included that of the 'Opening of the Dykes', a ritual following the annual flooding of the Nile in which the pharaoh himself would always cut the first irrigation channel.

In reality, however, much of the work was done for the pharaoh by his vizier, who was assisted by the various ministers, scribes and other officials who made up the Egyptian royal court.

GREAT ROYAL WIVES

Most pharaohs had several wives, although only one would be honoured with the title Great Royal Wife. To keep the bloodline pure, wives were often chosen from among close family members and commonly included sisters and half-sisters.

Female rulers

Some royal wives, such as Nefertiti, wielded considerable influence behind the throne, or upon the death of their husbands ruled the country until their children were old enough to assume their royal duties. Indeed, very rarely, as with Hatshepsut or Cleopatra, a woman became pharaoh in her own right. The first known female pharaoh was Sobekneferu, who is sometimes identified as the foster-mother of the biblical Moses.

MENES

TRADITIONAL FOUNDER OF DYNASTY I

Born: Possibly at Thinis, north of Abydos, exact date unknown. **Parents:** Unknown. **Reign:** c. 3050 BC (according to the historian Manetho, c. 60 years). **Wives:** Queens Berenib and Neithotepe. **Capital:** Memphis. **Achievements:** Unified all Egypt and founded Dynasty I. **Died:** Killed by a hippopotamus, date unknown (according to Manetho). **Tomb:** Possibly at Saqqara, Memphis. **Heir:** Djer.

UNIFIER OF EGYPT

Menes (otherwise known as Manes, Mni or Meni) is traditionally identified as the ruler who was responsible for the union of Upper and Lower Egypt around 3050 BC and as the first pharaoh of a united Egypt.

As well as founding the first dynasty he is also believed to have established the ancient cities of Crocodopolis and Memphis, the capital that he founded on an easily-

Warrior king

Menes is also said to have fought the Nubians in the south and to have extended the border as far south as Aswan.

defended island in the Nile south of modern Cairo. His other achievements included the development of commercial links with Byblos in Phoenicia.

SHROUDED IN LEGEND

The identity of Menes is a subject of debate and his life is shrouded in legend. Some historians believe that he may have been identical with Hor-Aha, a pharaoh who is said to have died when he was attacked by wild dogs and Nile crocodiles.

The Narmer Palette

The identification of the pharaoh was complicated even further in the nineteenth century by the discovery of the so-called Narmer Palette, a piece of slate bearing a relief depicting an early pharaoh of a united Egypt defeating his enemies in battle, but calling him Narmer (see page 19). Many Egyptologists now consider Narmer to have been the first pharaoh and Menes the second one.

Successor to the gods

Early accounts describe how King Menes dammed the Nile to reclaim land on which to build the city of Memphis. The same sources say that before Menes the throne of Egypt was held by a series of gods and demi-gods.

DJOSER

BUILDER OF THE FIRST PYRAMID

Born: Details unknown. **Parents:** Possibly King Khasekhemwy and Queen Nimaethap. **Reign:** c. 2668–c. 2649 BC (19 years, or possibly longer). **Wives:** Details unknown. **Capital:** Memphis. **Achievements:** Established ideal of kingship and built the first pyramid. **Died:** c. 2649 BC. **Tomb:** Step Pyramid at Saqqara, Memphis. **Heir:** Sekhemkhet.

Also called Netjerikhet Djoser or Zoser, Djoser was the second pharaoh of Dynasty III. The most notable event of his reign was a seven-year famine, which prompted him to erect a temple to Khnum (the god believed to control the flooding of the Nile) at Elephantine at Aswan. He was subsequently credited with ending the famine. It is thought that he extended Egypt's southern border and sent armies to subdue the Sinai. He is remembered for his six-stepped pyramid at Saqqara, near Memphis. Designed by Imhotep, it was the first pyramid and the earliest monumental building made of stone.

An ideal king

Djoser was greatly revered by later generations and considered an ideal of kingship by later dynasties.

SNEFERU

FOUNDER OF DYNASTY IV

Born: Details unknown. **Parents:** Possibly King Huni, unknown mother. **Reign:** c. 2613–c. 2589 BC. **Wife:** Queen Hetepheres. **Capital:** Memphis. **Achievements:** Built several pyramids and developed trade. **Died:** 2589 BC. **Tomb:** Possibly Red Pyramid at Dahshur. **Heir:** Khufu.

Sneferu is acclaimed as one of the wisest early pharaohs. His reputation rests on the major building projects he set in motion. His son Khufu may be remembered as the builder of the Great Pyramid of Giza, but Sneferu was even more active as a constructor of pyramids. He completed the step pyramid of Huni at Meidum, making it the first true pyramid (one with smooth, not stepped, sides), and also commissioned his own step pyramid at Meidum as well as the Bent Pyramid and Red Pyramid in the royal burial ground at Dahshur.

Seagoing pharaoh

Little is known about Sneferu's reign, but surviving records suggest he built up Egypt's fleets for military purposes and to develop trading links with Lebanon, the Sinai and Libya.

 Overleaf: The step pyramid of Djoser at Saqqara was the first of all the pyramids to be constructed.

KHUFU

BUILDER OF THE GREAT PYRAMID

Born: Details unknown. **Parents:** King Sneferu and Queen Hetepheres. **Reign:** c. 2589–c. 2566 BC (24 years). **Wives:** Queens Merityotes, Henutsen and one other. **Capital:** Memphis. **Achievements:** Organized building of the Great Pyramid of Giza. **Died:** c. 2566 BC. **Tomb:** Great Pyramid at Giza. **Heirs:** Radjedef, Khafra.

Called Cheops by the Greeks, Khufu succeeded his father Sneferu to become the second pharaoh of Dynasty IV, possibly when he was already advanced in years. He was among the most capable of the early pharaohs, centralizing the government of ancient Egypt and reducing the power of the priests.

A CRUEL TYRANT?

Khufu may also have sent military expeditions to the Sinai and ordered quarrying in the Nubian desert near Abu Simbel. Although little else is known about Khufu's reign, history (relying upon the writings of the Greek historian Herodotus) remembers him as a cruel tyrant. Herodotus claims that Khufu brought misery to both his family and his people, closing the temples and forcing men to labour in their thousands upon his pyramid-building projects.

When Hollywood films feature hordes of toiling workers being cruelly treated by overseers armed with whips as they slave in the hot deserts of ancient Egypt, building pyramids for the pharaoh, Khufu is the pharaoh who inspired such scenes.

THE PYRAMID OF KHUFU

The Great Pyramid at Giza was one of the famous Seven Wonders of the Ancient World as well as being the largest and oldest of all the pyramids that were constructed at Giza. It was built to house the body of King Khufu. Together with its temple complex, it took thousands of stonemasons and labourers more than 20 years to construct.

Pyramids for wives

Khufu also ordered the building of three smaller pyramids for each of his three wives. A special pit at the base of the Great Pyramid was also dug for his magnificent cedar wood funerary boat (which was later recovered by archaeologists).

A dynasty of pyramid builders

Khufu was succeeded (briefly) by his son Radjedef, and then by another of his sons, Khafra, the builder of the second largest pyramid at Giza.

KHAFRA

CREATOR OF THE SPHINX

Born: Details unknown. **Parents:** King Khufu and Queen Henutsen. Reign: c. 2558–c. 2532 BC (c. 26 years). **Wives:** Queens Meresankh II (his daughter-in-law) and Khameremebty I, among others. **Capital:** Memphis. **Achievements:** Built the Pyramid of Khafra at Giza and the Great Sphinx. **Died:** c. 2532 BC. **Tomb:** Pyramid of Khafra at Giza. **Heir:** Menkaura.

Khaf-Ra

The pharaoh's name is variously interpreted to mean 'Appearing like Ra' or 'Rise Ra!'

AN IMPORTANT BUILDER

Also known as Khafre, Chephren or Khephren, Khafra became the fourth pharaoh of Dynasty IV. He was a younger son of the pyramid builder Khufu and, like his

father, had a reputation for harshness and cruelty. He is remembered chiefly for his importance as a builder of some of the most famous of Egypt's ancient monuments.

Khafra ordered the building of the second-largest pyramid at Giza as well as the Great Sphinx, supposedly bearing his face and placed so as to guard both his own tomb and those of his father and step-brother. The associated temple is the only surviving temple dating from Dynasty IV.

For much of its life the Sphinx was buried up to its neck in sand, until it was excavated in the 1920s and 1930s.

MENTUHOTEP II

REUNIFIER OF EGYPT

Born: Details unknown. **Parents:** King Intef III and
Queen Iah. **Reign:** c. 2046–c. 1998 BC (51 years). **Wives:**
Queens Tem (his mother), Neferu (his sister), Henhenet
and others. **Capital:** Thebes. **Achievements:** Reunited
Upper and Lower Egypt and imposed administrative
reform. **Died:** c. 1998 BC. **Tomb:** Deir el-Bahri (near
modern Luxor). **Heir:** Mentuhotep III.

Nebhotepre Mentuhotep II (or Nebhepetra) was a
notable pharaoh of Dynasty XI. He was the son or
heir of Intef III and in due course he became the first
pharaoh of the Middle Kingdom period.

His reign probably began peacefully but was then
disrupted by military conflict. In the fourteenth year

 A painted sandstone statue of Mentuhotep II, wearing
the red crown of lower Egypt.

Amun-Re

When the two kingdoms of ancient Egypt were reunited
under Mentuhotep II their rival gods were combined in
a new divine figure, Amun-Re (or Amun-Ra).

of his reign he put down a revolt by the rival Dynasty X near Abydos, defeating the rebels in battle and driving his beaten enemies northwards. With his enemies repulsed he eventually succeeded in reuniting Upper and Lower Egypt, thus creating a united kingdom of Egypt for the first time since Dynasty VI.

CAPABLE IN WAR AND PEACE

Mentuhotep II is later thought to have led military expeditions south into Nubia, against an upstart Egyptian kingdom around Abu Simbel and, probably, into Palestine. He also restructured the administration of the country, which was placed under the control of a vizier, and by the time of his death he had brought peace and prosperity to the whole kingdom.

Building projects

Mentuhotep II constructed many temples and chapels in Upper Egypt. His most notable building project was his temple and tomb on the west bank at Thebes (now Luxor), in what later became known as the Valley of the Queens. This was the first complete mortuary complex known to have been constructed and may originally have included a pyramid or mastaba tomb as well as a terraced temple, avenues of statues and rows of pillars. The site was eventually excavated in 1968 but no royal sarcophagus belonging to Mentuhotep was found.

AMENEMHET I

WISE FOUNDER OF DYNASTY XII

Born: Details unknown. **Parents:** Senusret (a priest) and Nefret. **Reign:** c. 1991–c. 1962 BC (c. 30 years). **Wives:** Queens Nefrytatenen, Dedyet and Neferu. **Capital:** Itjtawy. **Achievements:** Restored prosperity and laid foundations for growth. **Died:** c. 1962 BC. **Tomb:** Pyramid at el-Lisht. **Heir:** Senusret I.

The reign of Amenemhet I (or Ammenemes I) marked the beginning of the Middle Kingdom of Egypt, when the country entered a period of recovery and prosperity. Though not of royal descent, Amenemhet I is thought to have served as vizier to his predecessor Mentuhotep IV before succeeding as pharaoh and suppressing rival claimants. He reasserted the authority of the throne and restored a tradition of pyramid-building associated with Dynasty VI. To ensure a smooth transition of power to his son, he shared the throne with him for 10 years (the first instance of such joint rule in Egyptian history).

Assassination

The circumstances of Amenemhet I's death are obscure, but it appears that he was one of the relatively few pharaohs to be assassinated, possibly by his own bodyguard.

SENUSRET I

PHARAOH OF DYNASTY XII

Born: Details unknown. **Parents:** King Amenemhet I and (possibly) Queen Nefrytatenen. **Reign:** c. 1971–c. 1926 BC (45 years). **Wives:** Queen Nefru. **Capital:** el-Lisht, near Memphis. **Achievements:** Expanded Egypt's borders and developed trading links. **Died:** c. 1926 BC. **Tomb:** Pyramid at el-Lisht. **Heir:** Amenemhet II.

Senusret I (or Sesostris I) became the second pharaoh of Dynasty XII on the assassination of Amenemhet I, having shared the throne as co-regent. He pursued military expansion into Nubia, extending the southern border of Egypt to the rapids on the Nile at Buhen, and pushing into the Libyan desert. He also established trade links with Syria and Canaan and offered support to local rulers who sided with him. He built temples and shrines at Karnak and two red granite obelisks at the rebuilt temple of Re-Atum in Heliopolis, one of which is now the oldest obelisk still standing in Egypt. He also rebuilt the Temple of Khenti-Amentiu Osiris at Abydos.

Trading success

Under Senusret I's stable rule, trade prospered and Egypt became increasingly wealthy.

SENUSRET III

WARRIOR-PHARAOH OF DYNASTY XII

Born: Details unknown. **Parent:** King Senusret II. **Reign:** c. 1878–c. 1841 BC (37 years). **Wives:** Queen Mereret and possibly his sister Sit-Hathor. **Capital:** el-Lisht, near Memphis. **Achievements:** Extended Egypt's southern border. **Died:** c. 1841 BC. **Tomb:** Pyramid at Dahshur or tomb complex south of Abydos. **Heir:** Amenemhet III.

The son of Senusret II, Senusret III (or Sesostris III) was the fifth pharaoh of Dynasty XII and one of the most powerful of the Middle Kingdom. He was revered as a warrior-king. Stability at home, by placing internal control under trusted viziers instead of local nobles, allowed him to focus on his military ambitions with repeated campaigns to extend and secure Egypt's southern border against the Nubians and protect trade routes. To maintain control of Nubia, he widened an existing canal at Aswan and built numerous fortresses. Plunder from his campaigns was used to decorate new buildings at Abydos and Medamud.

A king's responsibilities

Statues of Senusret III late in life depict him with a careworn expression, emphasizing his serious approach to his duties.

AMENEMHET III

SIXTH PHARAOH OF DYNASTY XII

Born: Details unknown. **Parent:** King Senusret III.
Reign: c. 1860–c. 1814 BC (c. 46 years). **Wives:** Queen
Aat and others. **Capital:** el-Lisht, near Memphis.
Achievements: Improved agriculture and continued
administrative reform. **Died:** c. 1814 BC. **Tomb:** Pyramid
at Hawara, in the Fayoum. **Heir:** Amenemhet IV.

Amenemhet III (also Ammenemes, Lamares, Ameres
or Moeris) may have shared the throne with his father
Senusret III for many years. Considered the greatest of
the Middle Kingdom pharaohs, he took advantage of a
long period of peace to promote the prosperity of the
Fayoum region of the Lower Nile, especially concerning
agriculture, and built barrages to aid land reclamation.
He also devoted much energy to building projects,
including the Temple of Sobek at Crocodopolis and
temples at Memphis and at Quban in Nubia.

The pharaoh's labyrinth

Amenemhet III's pyramid at Hawara incorporated some
of the most elaborate security features of any pyramid,
though it was still plundered by tomb-robbers in antiquity.
It may have inspired the labyrinth of King Minos in Crete.

SOBEKNEFERU

FIRST KNOWN FEMALE PHARAOH

Born: Details unknown. **Parent:** Possibly King
Amenemhet III. **Reign:** 1763–1759 BC (four years).
Capital: el-Lisht, near Memphis. **Achievements:**
First female pharaoh of Egypt. **Died:** 1759 BC.
Tomb: Unknown, possibly in a pyramid at Mazghuna.
Heir: Sekhemre Khutawy or Wegaf.

Sobekneferu (or Nefrusobek, Nefrusobk or Sobekkara)
may have been a daughter of Amenemhet III and was
the last pharaoh of Dynasty XII after Amenemhet IV
died early without an heir. The usual custom was for
descent to pass to males as the earthly representatives
of the male god Horus. It is possible that there were
earlier female pharaohs but no records exist. Though
said to be beautiful, Sobekneferu never married and
ruled without a king. Little is known of the events of her
brief reign, but she is believed to have commissioned
building at the tomb of Amenemhet III at Hawara. The
circumstances of her death and burial remain obscure.

Headless statues

Several statues of the queen have survived, although all of
them are headless.

HATSHEPSUT

FEMALE PHARAOH OF DYNASTY XVIII

Born: Details unknown. **Parents:** King Thutmose I and
Queen Ahmose. **Reign:** c. 1479–c. 1458 BC (22 years).
Husband: Thutmose II (half-brother). **Capital:** Thebes.
Achievements: Promoted trade and commissioned
buildings and monuments. **Died:** Disappeared, c. 1458
BC. **Tomb:** Unknown. **Heir:** Thutmose III.

Hatshepsut (or Hatchepsut) was the eldest daughter
of Thutmose I and Queen Ahmose and, following
tradition, was married to her half-brother Thutmose
II. When her husband died, the throne passed to the
pharaoh's young son Thutmose III and Hatshepsut
became the regent. She shared the throne with
Thutmose III for several years, but then she pushed
him aside and assumed sole authority as fifth pharaoh
of Dynasty XVIII.

ADOPTION OF MALE STATUS

Hatshepsut approached the role of pharaoh as the
equal of any man, dressing in male clothing, wearing
the ceremonial false beard of male pharaohs, adopting

A portrait bust of Hatshepsut, wearing the false beard
that was a symbol of kingship.

The first great woman

Hatshepsut is sometimes identified as the first queen to rule in her own right and as the first great woman in history. Her reputation soared after she was adopted as a feminist icon in the early twentieth century.

the male name Hatshepsu and taking a full part in the affairs of state. Her advisers included the high priest of Amun, Hapuseneb, and the royal steward, Senemut, who is popularly supposed to have been Hatshepsut's lover. Most notable of all her achievements was the re-establishment of neglected trade links, which were to bring great wealth to Egypt. Also significant were a trading expedition to the Land of Punt and military victories in Nubia, Syria and the Levant.

SHRINES DEFACED

Like her predecessors, Hatshepsut commissioned numerous buildings and statues. The most important building was the mortuary temple at Deir el-Bahri, designed by Senemut. Also notable were two obelisks erected at Karnak, at that time the tallest ever made.

Hatshepsut was ultimately overthrown when Thutmose III reclaimed the throne, and the exact details of her death are unknown.

THUTMOSE III

WARRIOR-PHARAOH OF DYNASTY XVIII

Born: Details unknown. **Parents:** King Thutmose II and Queen Isis. **Reign:** c. 1479–c. 1425 BC (54 years). **Wives:** Queens Hatshepsut-Merytre, Menhet, Menwi, Merti and others. **Capital:** Memphis. **Achievements:** Conducted numerous military campaigns, taking much plunder. **Died:** c. 1425 BC. **Tomb:** Valley of the Kings. **Heir:** Amenhotep II.

Thutmose III (otherwise known as Thutmoses or Tuthmosis III) was named as successor to his father Thutmose II, but on the latter's death power passed to his stepmother and aunt Hatshepsut, who acted as regent due to Thutmose III's youth.

Hatshepsut, however, later assumed the role of pharaoh for herself, and for over 20 years Thutmose was thrust into the background, perhaps sent to distant military commands. After Hatshepsut's death, Thutmose III reclaimed the throne as sixth pharaoh of Dynasty XVIII.

BATTLEFIELD COMMANDER

Thutmose III proved to be one of the great warrior-pharaohs of ancient Egypt and was a gifted battlefield commander. His many victories in battle were recorded on the walls of the temple at Karnak. He conducted numerous campaigns in the Levant, in the course of which he captured 350 cities.

His triumphs included the capture of the Canaanite city of Megiddo after a battle and siege. He also became the first pharaoh to lead his armies over the River Euphrates. Considerable spoils from these campaigns were sent back to Egypt. Every summer for 18 years he also embarked on military campaigns against Syria, making Egypt the dominant force in Palestine.

BUILDING PROJECTS

Thutmose III commissioned many notable building projects, including his own temple which was near that of Hatshepsut. Most celebrated of all were his additions to the temple complex at Karnak, to which he presented many gifts of plunder. His apparent destruction of monuments commemorating Hatshepsut suggest his relationship with her was difficult, although he may have just been erasing memory of a female pharaoh (then a controversial idea).

Military genius

Because of his brilliance as a military strategist, Thutmose III is sometimes called the 'Napoleon of Egypt'.

AMENHOTEP III

PHARAOH OF DYNASTY XVIII

Born: c. 1395 BC. **Parents:** Thutmose IV and Queen Mutemwiya. **Reign:** c. 1389–c. 1351 BC (39 years). **Wives:** Queen Tiy and others. **Capital:** Thebes. **Achievements:** Presided over a period of peace and prosperity as well as artistic achievement. **Died:** c. 1351 BC. **Tomb:** Western Valley, Valley of the Kings. **Heir:** Akhenaten.

The grandson of the warrior-pharaoh Thutmose III, Amenhotep III succeeded Thutmose IV while he was still a child, and his mother Mutemwiya probably ruled in his name in his early years.

Egypt was prosperous and stable under Amenhotep III and his reign witnessed a peak in Egyptian civilization. Peace with Egypt's neighbours left the pharaoh free to concentrate upon developing diplomatic and trading links and promoting cultural achievements from the

Poor health

Images of Amenhotep III suggest he was relatively sickly. It is clear from his mummy that he probably suffered badly from toothache in his last years.

comfort of a luxurious court. Notable building projects included additions to the temple at Luxor, rebuilding at Karnak and a massive mortuary temple at Thebes, of which all that remains are an impressive pair of giant statues, one depicting Amenhotep III himself and the other his wife and mother.

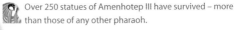

Over 250 statues of Amenhotep III have survived – more than those of any other pharaoh.

AKHENATEN

THE HERETICAL PHARAOH

Born: c. 1379 BC. **Parents:** King Amenhotep III and Queen Tiy. **Reign:** c. 1353–c. 1336 BC (17 years). **Wives:** Queens Nefertiti, Merytaten, Kiya, Mekytaten and Ankhesenpaaten. **Capital:** Akhetaten. **Achievements:** Introduced short-lived cult of Aten. **Died:** c. 1336 BC. **Tomb:** (Initially) Royal Tomb at Amarana; probably Valley of the Kings. **Heirs:** Smenkhkare, Tutankhamun.

After the premature death of Amenhotep III's eldest son Thutmose the throne passed to a younger son, who became the tenth pharaoh of Dynasty XVIII, initially as Amenhotep IV. During the first year of his reign, however, the new pharaoh changed his name to Akhenaten (otherwise Akhnaten, Akhenaton, Akhnaton, Ankhenaten, Ankhenaton or Ikhnaton), meaning 'He Who is of Service to Aten'.

Erased from history

Later pharaohs accused Akhenaten of heresy and erased him from history, and it was not until the nineteenth century that archaeologists finally restored his reputation. He is now considered to be among the most remarkable of all the pharaohs.

His wife Nefertiti, meanwhile, was restyled Nefer-Nefru-Aten, meaning 'Beautiful is the Beauty of Aten'. These name changes reflected a major shift in Egyptian religion and culture resulting from the pharaoh's controversial decision to replace Amun-Re as the supreme deity with the hitherto obscure sun god Aten, represented by a solar disc.

CULT OF THE SUN GOD

The reasons for Akhenaten's extraordinary rejection of the old religion have been much debated. Initially the new god was worshipped alongside the old ones, with a temple to Aten being erected alongside that of Amun-Re at Karnak, but Akhenaten then decreed that Aten was the only deity and that only he, Akhenaten, could communicate with him. The temples of the other gods were closed, the powerful independent priesthood was dismantled and the seat of government was moved to a new capital city called Akhetaten (modern el-Amarna). Any statues and shrines to Amun-Re were destroyed or defaced.

Consequences of sun worship

In response to the new religion came a distinctive and naturalistic new style of art, which was largely inspired by the striking appearance of the pharaoh Akhenaten himself, featuring elongated faces, slender limbs and protruding stomachs.

Akhenaten's interest in religion and philosophy meant that he left the other affairs of state to his advisers and his reign saw a gradual decline in Egypt's national fortunes, with increasing corruption and disorder. The cult of Aten did not long survive Akhenaten's death and the city and temples he had built were dismantled or allowed to fall into ruin.

 The accuracy of depictions of Akhenaten remains speculative as his mummy has never been found.

TUTANKHAMUN

THE BOY PHARAOH

Born: c. 1342 BC. **Parents:** Possibly King Amenhotep III and one of his wives or, more likely, King Akhenaten and Queen Kiya. **Reign:** c. 1334–c. 1325 BC (nine years). **Wives:** Queen Ankhesenpaaten/Ankhesenamun. **Capital:** Thebes. **Achievements:** Began restoration of cult of Amun. **Died:** c. 1325 BC. **Tomb:** Valley of the Kings. **Heir:** Ay.

Tutankhamun (or Tutankhamen) is perhaps the most famous of all the pharaohs, though in reality his reign was not especially noteworthy. At the age of nine he succeeded the short-lived Smenkhkare as the twelfth pharaoh of Dynasty XVIII, initially as Tutankhaten

Death of a pharaoh

Tutankhamun is thought to have died around the age of 19 years. A recent examination of his mummy suggests that he died of gangrene after severely breaking his leg, possibly in a chariot accident. His tomb was also found to contain the mummies of two stillborn daughters.

 The young pharaoh and his queen, depicted on the magnificent golden throne found in his tomb.

A gold statue of the boy pharaoh hunting hippopotamuses.

(meaning 'Living Image of Aten') but changing his name to Tutankhamun (meaning 'Living Image of Amun') as a gesture of his rejection of the cult of Aten which had made the reign of his predecessor Akhenaten controversial. The priests of Amun had their privileges restored, worship of the old gods was once more permitted, and the capital of Egypt was returned to Thebes. Due to Tutankhamun's youth, the affairs of state were probably handled chiefly by his vizier Ay.

KING TUT

The young pharaoh's modern fame owes everything to the discovery of his tomb, remarkably intact, in the Valley of the Kings in 1922. This stunning discovery inspired worldwide interest in the civilization of ancient Egypt and over the years millions of people have seen the priceless artefacts with which the boy pharaoh was buried.

SETI I

PHARAOH OF DYNASTY XIX

Born: c. 1319 BC. **Parents:** King Ramesses I and Queen Sitre. **Reign:** c. 1291–c. 1279 BC (c. 13 years). **Wives:** Queen Tuya and others. **Capital:** Memphis. **Achievements:** Reconquered territories lost to the Hittites and promoted art and culture. **Died:** c. 1279 BC. **Tomb:** Valley of the Kings. **Heir:** Ramesses II.

Named in honour of the god Set, but also known as Sethosis or Menmaatre (meaning 'Eternal is the Justice of Re'), Seti ruled initially alongside Ramesses I, overseeing Egypt's military operations while his father took care of domestic affairs. One of the great pharaohs, as second pharaoh of Dynasty XIX, he asserted Egyptian control of Canaan and Syria against the Hittites, winning important victories and reclaiming much lost territory. Scenes from his battles were mounted on the walls of the temple of Amun at Karnak. Major building projects commissioned by Seti included the hypostyle hall at Karnak and fine temples at Thebes and Abydos.

Sudden death

Examination of the mummy of Seti I revealed that he died suddenly around the age of 40, possibly from heart disease.

RAMESSES II

RAMESSES THE GREAT

Born: c. 1302 BC. **Parents:** King Seti I and Queen Tuya. **Reign:** c. 1279–c. 1213 BC (c. 66 years). **Wives:** Queens Nefertari, Istnofret, Binthanath, Merytamon, Maathornefrure among many other wives and concubines. **Capital:** Pi-Ramesse (near modern el-Khatana). **Achievements:** Achieved many military victories and built some of Egypt's most famous monuments. **Died:** c. 1213 BC. **Tomb:** (Initially) Valley of the Kings; later moved to Deir el-Bahri. **Heir:** Merneptah.

Ramesses II (otherwise Rameses or Ramses) probably shared the throne with his father before becoming the third pharaoh of Dynasty XVIV in his early twenties. Alongside his father he took part in military campaigns in Libya and Nubia. After Seti's death he waged war on the Syrians, against whom he fought repeated campaigns. His most famous battles included the Second Battle of Kadesh in 1274 BC, in which he confronted the Hittites under King Muwatallis and effectively prevented further assaults by enemy forces. Peace was eventually agreed with the Hittites after it became evident that neither side could achieve a final victory over the other.

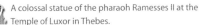

A colossal statue of the pharaoh Ramesses II at the Temple of Luxor in Thebes.

The biblical pharaoh

Ramesses II is traditionally identified as the pharaoh who was on the throne of Egypt at the time of Moses and the biblical Exodus, though there is little evidence for this.

ABU SIMBEL

The reign of Ramesses II saw the construction of some of Egypt's most celebrated ancient sites. These included two temples at Abu Simbel, dominated by images of Ramesses himself, the completion of the famous hypostyle hall at Karnak begun by Seti I, the mortuary complex near Luxor known as the Ramesseum and the Colossus of Ramesses at Memphis (a massive statue of the pharaoh, now fallen). Ramesses is reputed to have commissioned more monuments and statues than any other pharaoh and numerous examples have survived.

Ramesses' mummy

Thought to have been around 90 years old when he died (although he was probably not 99 as legend claims), Ramesses outlived the first 13 of his heirs. He had many sons and daughters by his numerous wives (whose number is said to have been around 200). His mummy was among those found in the royal tomb unearthed at Deir el-Bahri in 1881; it is now housed in the Egyptian Museum in Cairo.

 A depiction of Ramesses II striking down Libyan and Nubian warriors.

PTOLEMY I SOTER

FOUNDER OF THE PTOLEMAIC DYNASTY

Born: 367 BC. **Parents:** Lagus or Philip II of Macedon and Arsinoe of Macedonia. **Reign:** 323–285 BC (38 years). **Wives:** Princess Artacama of Persia, Queens Thaïs and Berenice. **Capital:** Memphis. **Achievements:** Achieved numerous military victories and extended Egyptian possessions in the eastern Mediterranean. **Died:** 283 BC. **Tomb:** Unknown. **Heir:** Ptolemy II Philadelphus.

Ptolemy I Soter was a veteran Macedonian general who assumed control of Egypt in 323 BC on the death of Alexander the Great, to whom he had given many years of loyal service. There were several other claimants to Alexander's possessions and to secure Egypt Ptolemy had first to crush the armies of his rivals, notably Alexander's cavalry commander Perdiccas.

Claiming the throne

According to Macedonian custom, the heir to a dead king asserted his claim to the throne by burying his predecessor. Ptolemy went to great lengths to secure the body of his friend Alexander the Great so that he could bury it before any of his rivals did so. The site of Alexander's grave is thought to be somewhere in modern Alexandria.

A GOOD MILITARY COMMANDER

After many years of war, Ptolemy eventually secured his hold on Egypt and some neighbouring territories, such as Syria and Cyprus. He even seized parts of Greece. In 305 BC, he finally accepted the title of pharaoh, as Ptolemy I, thereby founding a great pharaonic dynasty that was to last for 300 years.

A SHREWD RULER

A shrewd man who maintained order at home, Ptolemy amassed huge wealth as well as making significant contributions to Egyptian culture, founding the Great Library at Alexandria and writing an account of Alexander's campaigns. The few major monuments that were commissioned during his long reign included the temple of Kom Abu Billo which was dedicated to the goddess Hathor. Ptolemy I Soter abdicated in 285 BC, when he was in his eighties, in favour of one of his sons, who had shared power with him for three years before assuming the throne as Ptolemy II Philadelphus.

Saviour of Rhodes

Ptolemy was dubbed Soter (meaning 'Saviour') after sending help to Rhodes when it was being besieged in 305–304 BC by Demetrius.

CLEOPATRA VII

THE LAST OF THE PHARAOHS

Born: Alexandria, 69 BC. **Parents:** King Ptolemy XII
Auletes. **Reign:** 51–30 BC (21 years). **Husbands:**
Ptolemy XIII and Ptolemy XIV (her brothers). **Capital:**
Alexandria. **Achievements:** She sought to maintain
Roman support for Egypt and acquired semi-legendary
status herself. **Died:** Committed suicide, 30 BC.
Tomb: Unknown.

When Ptolemy XII Auletes died in 51 BC he left the
throne jointly to his daughter Cleopatra (then 18) and
her younger brother Ptolemy XIII (then 12). Passionate,
ambitious and clever (she was the only one of the
Ptolemies to learn Egyptian), Cleopatra immediately
showed signs of independence, refusing to defer to her
younger brother and dreaming of rebuilding Egypt's
empire. In 48 BC, however, Cleopatra's enemies at court
engineered her overthrow in favour of Ptolemy and
she had to flee Alexandria.

The legend

One of the most colourful and well-known figures of
the ancient world, Cleopatra has been the inspiration
of countless paintings, plays, films and other works.

 The story of Cleopatra's reign took on the aura of romantic myth to later generations of artist and writers.

The Ptolemaic dynasty relied heavily on the support of Rome, especially that of Pompey. When, however, Pompey was defeated by Julius Caesar at Pharsalus, Ptolemy sought Caesar's favour by having Pompey murdered. Unfortunately for Ptolemy, Caesar was outraged by this treachery and occupied Alexandria.

Goddess of love

Cleopatra greatly impressed Antony at their first meeting, making a grand waterborne entrance dressed as Aphrodite, the Greek goddess of love.

Cleopatra seized the chance to win Caesar over by famously having herself secretly delivered to him rolled up in a carpet (or, possibly, a sack). She and Caesar became lovers, which infuriated Ptolemy, whose soldiers then surrounded Alexandria. Caesar killed several of Ptolemy's men and Ptolemy himself was drowned crossing the Nile.

CAESAR AND CLEOPATRA

Cleopatra now claimed the throne for herself. She married her brother Ptolemy XIV to win over the Egyptian priesthood but continued her relationship with Julius Caesar. Caesar and Cleopatra sailed up the Nile together to Dendera, where she was worshipped as pharaoh. Their son Caesarion was born shortly afterwards and in 46 BC Cleopatra accompanied Caesar to Rome. Cleopatra's presence in Rome was hugely controversial, as were Caesar's plans to marry her. In 44 BC Caesar was assassinated and Cleopatra hastened back to Alexandria, fearing that she and her son would also be murdered.

ANTONY AND CLEOPATRA

Ptolemy XIV died in mysterious circumstances and Caesarion was made co-regent. When Mark Antony emerged as the leading Roman commander, Cleopatra set about seducing him and winning him over. Antony and Cleopatra spent the winter of 41–40 BC together in Alexandria. In 40 BC, however, Antony returned to Rome and made peace with Caesar's legal heir Octavian by marrying Octavian's sister Octavia. Cleopatra, meanwhile, gave birth to twins (a boy and a girl) by Antony.

Antony and Cleopatra were reunited at Antioch, where Antony presented her with much-needed land. With Egyptian support, he attacked the Parthians, but was defeated in 36 BC. A third child was born and, his army in tatters, he returned to Alexandria with Cleopatra.

When he divorced Octavia, Octavian declared war against Cleopatra. In 31 BC Octavian's navy defeated Antony at Actium in Greece and within six months Octavian's soldiers had reached Alexandria itself. Antony offered feeble resistance and committed suicide by falling on his sword. To escape being displayed as Octavian's slave Cleopatra had an asp (an Egyptian cobra) brought to her concealed in a basket of figs. She died of snakebite on 12 August 30 BC, aged 39. Caesarion was strangled to death.

PART FIVE

Life in ancient Egypt

We know a surprising amount of detail about the lives that were led by ancient Egyptians. Hieroglyphics, tomb paintings and archaeological evidence have all enabled modern historians to learn much about the society the Egyptians lived in, as well as providing fascinating information about their jobs, their houses, their families, their diet and many other aspects of everyday life.

Leisure

In prosperous times, the wealth of Egypt allowed many members of society to enjoy themselves pursuing various leisure interests.

EGYPTIAN SOCIETY

Egyptian society consisted of three broad classes: the upper, middle and lower classes. At their head was the pharaoh and the royal family, who held ultimate power. Making up the rest of the upper class were the viziers, priests, scribes, physicians and various noblemen and officials of the royal court.

Traders and merchants belonged to the middle class, upon which Egypt's great wealth largely depended. The middle class also included craftsmen, such as carpenters and jewellers, and dancers.

The largest group was the lower class, which was made up of peasant farmers and other unskilled labourers. When times were good, members of the lower class shared in the country's prosperity, but when times were bad they were the first to suffer.

THE UNDERCLASS

At the foot of society was an underclass of servants and slaves. This group included the maids, gardeners and cooks who were employed in the households of wealthy families, as well as captured enemy soldiers who were forced to work on the pharaoh's monuments and other construction projects.

THE UPPER CLASS

Members of Egypt's upper class enjoyed the best standard of living. The nobility owned the largest houses, wore the finest clothes and jewellery and ate the choicest food. Their wealth meant that they could also keep servants and slaves to perform routine household tasks for them. When they died, they alone could afford the expense of having their bodies properly mummified and laid to rest in richly decorated coffins, together with valuable grave goods.

Education

Only the children of the upper class were educated and could take on the important jobs in society as adults. This meant that the noble families tended to hold onto their aristocratic status through many generations.

Viziers

After the pharaoh and his family, the most important members of the upper class were the two viziers, who ruled Upper and Lower Egypt on behalf of the pharaoh. Some viziers, such as Tutankhamun's vizier Ay, even went on to become pharaoh themselves.

Each vizier had a number of royal overseers under their command, each with their own responsibility, such as the army or the country's granaries. Viziers were also judges and ran the administration of the country.

Meeting the pharaoh

If invited to approach the pharaoh, courtiers had to kiss the ground beneath his feet as a gesture of their humility. Those who enjoyed the pharaoh's particular favour might be allowed to kiss his leg instead.

LIFE IN THE ROYAL COURT

The most privileged members of the upper class were allowed access to the court of the pharaoh himself. Here they could enjoy great banquets and be waited upon by servants in splendid surroundings. The first duty of courtiers was to satisfy the daily demands of the pharaoh and members of his family, in whom ultimate power lay.

Ceremonial occasions

The pharaoh held court under a canopy, attended by fan bearers and sitting on a golden throne, similar to those found in the tombs of Queen Hetepheres and Tutankhamun. The court comprised a large number of officials, including viziers, royal bodyguards, lawyers, tax officials, priests, architects and army officers. Surrounded by his court, the pharaoh would celebrate major ceremonial occasions, such as coronations, royal jubilees or receptions for foreign diplomats.

An example of a royal fan of beaten gold, bearing cartouches with the name of the ruling pharaoh. Such fans were symbols of the king's power, and it was a great honour to be appointed a royal fan bearer.

THE MIDDLE CLASS

Below the upper class were the traders and merchants who made up Egypt's middle class, and whose efforts were the main source of the country's great wealth. The pharaohs insisted that taxes be paid on many of the goods that were traded and these contributed hugely to the royal treasuries. The ranks of the middle class included craftsmen of many kinds who lived with their families above the shop where they worked. Finished products might be bartered in local markets.

Merchants and traders

Traders did not use money but bartered (exchanged) goods. They also exchanged their products with the traders in neighbouring countries and carried them by boat to be traded with more distant peoples. Some merchants became very rich from their business and used their wealth to enjoy lifestyles that were similar to those of the upper class. However, others operated on a smaller scale and were only marginally better off than the lower class.

Family businesses

The children of craftsmen usually followed into the same trade as their fathers, learning by becoming assistants in the family workshop.

THE LOWER CLASS

Ancient Egyptians belonging to the country's lower class lacked any formal education or skills beyond those of working in the fields. The majority of the population was engaged in farming. When they were not busy in the fields these agricultural labourers might be ordered by the pharaoh to work on the construction of the pyramids or other major monuments.

Everyday life

The poorest members of Egyptian society lived in small, often overcrowded, houses and they had few luxuries in their lives. During the day most men would be out of their villages, toiling in the fields, while their women remained at home weaving or looking after children. Their diet was simple and meals were often no more than bread and onions washed down with beer.

Overleaf: A wall painting from Giza depicting a variety of everyday activities in ancient Egypt, including farming and men playing the senet board game.

Military service

In times of war, in the years before the establishment of a proper army, ordinary Egyptians would also be called upon to fight off any invaders.

Work gangs

Many unskilled workers were required to work on building the massive tombs for the pharaohs and other monuments. These labourers were organized into gangs under the control of a foreman, who directed the work assisted by a scribe. Any quarrels between these workers would be settled in a local village court comprising the foremen, scribes and other prominent village officials.

 This is a carving of a Nubian slave, who may have been captured by the Egyptian army.

SERVANTS AND SLAVES

Wealthy Egyptian families could afford several servants to perform domestic jobs around the home, including cooks, gardeners and maids. The richest families even had servants to look after their wigs. Family servants were often given their own rooms within the family home, separate from those of family members. Rich Egyptians wanted to be sure of having servants in the afterlife and so, when they died, had models of servants (called shabtis) buried with them in the belief that these would come to life to satisfy their needs.

SLAVES OF THE PHARAOHS

Soldiers captured by Egypt's armies were often taken home to serve as slaves of the pharaohs, working on temples and tombs and other great monuments. The life of an Egyptian slave was not always as bad as popularly supposed. Slaves were allowed to own land and other property, and could even buy their freedom.

Servants of the dead

When Tutankhamun's tomb was opened it was found to contain no less than 236 shabti figures to serve the dead king in the afterlife.

TOWNS AND VILLAGES

Most ancient Egyptians lived in towns and villages on the banks of the Nile or in the countryside on estates belonging to high officials at the pharaoh's court. In Egypt's early history the only major towns were the capital cities of Memphis and Thebes, but large settlements grew up elsewhere over the centuries.

Settlements sprang up near oases or wells where fresh water could be easily obtained, with buildings lining either side of narrow streets. The various buildings might include grain stores, workplaces for craftsmen and temple complexes as well as private houses. Apart from temples and palaces, most buildings were made of mud brick and had flat roofs where people spent much of their time, escaping the heat inside their rooms. Other features might include a marketplace.

ARCHAEOLOGICAL REMAINS

The remains of ancient Egyptian settlements may still be seen. The best-known of these include the village of Deir el Medina, which was founded in the sixteenth century BC and was home to around 60 families over a period of some 500 years. The village was built in the desert near the Valley of the Kings for the workers building tombs for the pharaohs, and, as elsewhere, the

White houses

Most houses in ancient Egypt were painted white in order to reflect the heat of the strong sun.

houses were closely packed. Typical homes were long and narrow in shape, with a relatively short street frontage. Most had around four or five rooms, with the kitchen in a yard at the back.

HOUSES

Ordinary ancient Egyptians built their houses from mud bricks, stone being expensive and thus reserved for temples and other important buildings. Because of this, few traces remain of ordinary homes, which were swept away in floods over the centuries. Mud bricks were made by trampling river mud underfoot and mixing into it pieces of straw and pebbles. The bricks were then shaped and left to dry in the sun until they were ready for use. The inside walls of the finished house were coated with lime plaster and might then be painted with murals. In the case of traders and craftsmen, the lower storey of the house might be reserved for business, while the upper storeys provided living accommodation.

In order to keep houses cool, there were only a few small windows set high up in the walls. The most comfortable place in the house was the flat roof, where the family would eat, talk and play board games. Vents on the roof directed cool north breezes into the interior, which was lit by oil lamps at night. The furnishings included beds, short-legged wooden chairs and tables, footstools, chests and storage baskets as well as ornaments and shrines. The floors might be covered with mats of woven reeds.

WEALTHIER HOMES

While the poorest families might share a single-roomed house, wealthy people had larger homes or villas with several storeys, containing rooms for both family members and servants, including a bathroom and lavatory. The grandest houses might have large porches supported by rows of columns with tops shaped like lotus buds. The kitchens would be sited away from the other living quarters, so that cooking smells would not spread through the whole house. Outside might be a pleasant walled garden, planted with trees and shrubs, and a pool stocked with lotuses and fish.

Sleeping quarters

Ancient Egyptians slept on wooden-framed beds of canvas or rushes. They placed their heads on wooden rests, which allowed cool air to circulate around them as they slept.

FAMILY LIFE

Although little remains of the houses in which ancient Egyptians lived, many details of their domestic lives can be learned both from wall paintings and objects that have been found in tombs.

It seems that most men had just one wife, although pharaohs and some more important figures in society might have had several wives. Although considered secondary to their husbands, women enjoyed the right to own land and property, including the furniture in their homes. If the family was wealthy enough, they might have several servants or slaves to perform routine domestic tasks, such as cooking and cleaning, or to carry their masters on litters from place to place.

CHILDREN AND PETS

Children of ordinary families divided their time between sleeping, playing, helping with household tasks and, when older, working alongside their parents.

Pets in the afterlife

When pets died they might be mummified. Pet dogs were often buried with their collars for use in the afterlife.

If their parents were wealthy, boys were sent to the temple school from around the age of nine to learn how to read and write, but other children received no formal education. Treats included feasts on special occasions and hunting trips with the rest of the family.

Egyptian families liked to keep pets, such as dogs, cats (which they called 'miws'), doves and monkeys. The Egyptians are believed to have been the first to tame cats, originally to catch mice in the grain stores.

 Cats were popular pets in ancient Egypt. They kept vermin under control and were sometimes used to help in hunting wild birds.

FOOD

The Egyptians were good herdsmen and farmers and, as a result, except in times of famine when the Nile flood failed, all but the poorest people enjoyed a varied diet of meat, fish, fruit and vegetables. In order to keep food from going rotten in the hot climate, it was often rubbed with salt or dried.

BREAD AND STAPLES

Wheat and barley were used to make bread, the staple food of ancient Egypt. The dough was trodden with the feet in a huge tub, but the resulting bread was not of the highest quality as it contained a lot of grit. Many of the mummies found by modern archaeologists have been found to have cracked and worn-down teeth from eating such bread. Egyptian bakers also made ring doughnuts, buns shaped like pyramids and cakes shaped like crocodiles.

Food for the poor

Poor people ate less well than the rich, who could afford a range of meats and other dishes. A typical lunch eaten by a worker in the fields comprised barley bread and onions or cucumber, washed down with beer. Ordinary people also ate porridge made from barley flour, as well as fish, vegetables, garlic and dates.

DRINK

Barley was used to make beer, the favourite drink of ordinary ancient Egyptians. This beer was very thick and had to be strained before drinking. The Egyptians also knew how to tread grapes to make wine, although this was drunk only by the wealthier members of society. Wine was also made with dates.

BANQUETS

Wealthy Egyptians prepared banquets for their friends and might spend a whole day cooking a range of rich dishes to be served up. Meat dishes provided on such occasions could include beef, duck, goose, gazelle, heron, mutton, quails and veal. These meats might be barbecued or roasted with onions, garlic, spinach, leeks, peas, beans, lentils, beans, fish or other foods. Guests would then be offered fresh fruit, such as figs, melon and pomegranates, as well as cakes and pastries sweetened with honey. Other delicacies included lotus root and cheese. The usual drink was wine.

Food for the dead

Food was often left in tombs for the dead to eat. Some tombs also contained a model of a brewer to provide the deceased with beer in the afterlife.

Duties of a host

Banquets were usually served under cover on the flat roof of the host's house, where it was cooler. Everyone wore their best clothes, wigs and jewellery. Hosts and guests of honour sat on low chairs, while others sat on mats. The food itself was placed on low tables. Music and other entertainment, such as acrobatics, might be organised to complete the occasion.

Finger food

Guests did not use cutlery but instead ate with their fingers, which they then washed in dishes of scented water which were brought to them by servants.

 Wall painting depicting Egyptian farmworkers picking grapes to be trodden and made into wine.

CLOTHING AND HAIRSTYLES

Because of the hot climate of north-east Africa, ancient Egyptian adults wore very little in the way of clothing, and their children usually nothing at all.

MEN'S AND WOMEN'S CLOTHING

Most ordinary men wore simple loin cloths around the waist or short kilt-like skirts and little else, although more important men might wear longer, calf-length pleated skirts of fine linen (made from flax).

Egyptian women wore long sheath-like dresses with shoulder straps, sometimes with beads sewn into them, and light pleated linen cloaks over the top.

Most clothing was plain white in colour, in order to reflect the sun's heat, although it was sometimes dyed bright orange, green or yellow. Senior priests wore leopard skins as a mark of their status in society.

Footwear

Poor people went about barefoot or wore simple sandals made from papyrus or other reeds. Rich Egyptians might have sandals made of leather or even gold.

WIGS

All Egyptians shaved their heads, using bronze razors as scissors had yet to be invented. Rich Egyptians, including the pharaoh himself, covered their bare heads with black shoulder-length wigs. These wigs were made either from human hair or sheep's wool, which was woven into plaits and then held in place with some beeswax.

Young girls wore pigtails, while young boys had their heads shaved, with a single plait of hair left dangling on the right-hand side of the head as a symbol of their youth. Wigs could be curled using tongs or kept tidy with pins and combs, examples of which have been found in many tombs.

THE ROYAL WARDROBE

The pharaoh himself wore garments which were made from the finest linen available. He is often depicted wearing a kilt-like skirt which was wrapped around the hips and then tied at the waist. This was sometimes richly decorated and colourful.

Over his wig, the pharaoh might wear a blue-and-white striped linen headdress called a *nemes* or a single-coloured headcloth called a *khat*. This was replaced by a crown on ceremonial occasions.

MAKE-UP

Both Egyptian men and women thought that their personal appearance was very important and as well as choosing fine clothes they wore lots of make-up, which they made from finely ground minerals.

Rich men and women drew a distinctive heavy black line round their eyes, using an eyepaint known as kohl, which was made from a lead ore called galena. They might also wear green eyeshadow which was made from a copper ore called malachite.

The women also applied rosy lipstick as well as cheek blusher which was made from iron oxide, painted their fingernails and kept a store of cosmetic creams, cosmetic holders, make-up applicators and powders, many examples of which have since been found in tombs. Mirrors and combs were commonly used by Egyptians to perfect the appearance.

Medicinal make-up

The black line that Egyptians painted round their eyes helped to reduce the glare of the sun and also contained a mild disinfectant which was useful in protecting the eyes from diseases spread by flies.

SWEET-SMELLING

When going to a feast, guests pinned cones of animal fat scented with herbs and spices to their hair. As Egypt is a hot country, the fat slowly melted as the evening passed, giving the guests sweet-smelling (but greasy) hair and skin. They might also apply perfumed oils as deodorant, use breath fresheners and carry sweet-smelling flowers. The Egyptians also used special lotions that were supposed to prevent baldness, dandruff and spots.

 The ancient Egyptians took great care over their appearance, wearing wigs, jewellery and make-up.

JEWELLERY

Wealthy Egyptian men and women wore jewellery both for personal adornment and as a mark of their status. The best jewellery was made of gold, which was mined in the desert east of the Nile and fashioned into intricate designs. On special occasions rich men wore fabulous bracelets and wide golden necklaces or collars studded with precious stones. On their chests they might also wear a large piece of jewellery called a pectoral, which might be shaped in the form of a falcon, scarab beetle or similar sacred image.

As well as rings and earrings, wealthy women wore beautiful necklaces which were decorated with gold or ceramic ornaments, as well as elaborate bracelets, girdles and anklets. Poorer people wore copper jewellery adorned with beads or shells.

LUCKY AMULETS

Also popular with all Egyptian classes were various amulets (lucky charms), such as the eye of Horus, that were supposed to ward off evil influences. Children might be given special fish amulets that were believed to prevent them drowning in the river Nile, while luck-preserving amulets were also placed in the linen bandages of mummies to protect the dead.

Gems

Diamonds, emeralds and rubies were unknown to the ancient Egyptians, but their jewellers made extensive use of other gems that were available. These included cornelian (red), amethyst (mauve), turquoise (blue-green), lapis lazuli (blue) and feldspar (green).

The colour of gems was very significant to the Egyptians. The ruby red colour of red jasper, for instance, represented blood and life, and was thus used in the girdle of Isis amulets which were placed on the throat of mummies. The blue of lapis lazuli symbolized the sky, while the green of malachite, turquoise, green feldspar, serpentine and other gems represented new growth and fertility.

EGYPTIAN JEWELLERS

Although many items of jewellery arrived in Egypt as tribute or as spoils of war from Syria and the north, much more was the handiwork of native Egyptian jewellers, who laboured in workshops which were attached to temples and palaces.

The names of officials presiding over these jewellers are known, but those of the craftsmen are obscure. They could be sub-divided into those who worked as goldsmiths (*neby*) or as makers of jewellery (*neshdy*).

LEISURE ACTIVITIES

The Egyptians had many ways of spending their spare time. As well as physical activities, such as ball games, acrobatics and swimming in the Nile, they also enjoyed feasting, music and dancing and intellectual pursuits.

BOARD GAMES

They devised a number of popular board games, of which the most famous was senet, a game a little like modern chess or ludo. It was played with a set of playing pieces made of white ivory and black ebony wood, which were moved according to how the player's throwsticks (the equivalent of modern dice) landed. The game supposedly represented the battle between good and evil on the journey to the underworld. Another board game was snake, in which two players raced each other's circular stone counters (like marbles) round a spiral board, the winner being the first to reach the middle. Other popular board games included one called hounds and jackals.

CHILDREN'S GAMES

Children played with a variety of toys, which included wooden lions and other carved animals (many with moving parts), spinning tops, clay balls that rattled and

The game of kings

Archaeologists have found many senet boards and pieces, including no less than four sets that were buried with the boy-pharaoh Tutankhamun to amuse him in the afterlife.

dolls with beads in their hair. Popular games included leapfrog and tug-o'-war. Adults also entertained their children with tales about gods and goddesses.

MUSIC AND DANCING

Wealthy people would regularly entertain their friends at parties and feasts, offering their guests a variety of entertainment, which usually consisted of music, dancing and acrobatics.

Depictions of musicians and dancers in wall paintings suggest that the dancing was done chiefly by scantily clad dancing girls, while the musicians sang or played their harps, lutes, drums, reed pipes, double oboes, tambourines, bone castanets and cymbals. Many examples of these instruments have survived to this day, although, unfortunately, there is no record of the music that was played.

 Overleaf: Musicians, wearing sweet-smelling cones of scented fat on their heads, play for dancers at a feast.

Egyptian sand dance

Nowadays people often imagine the ancient Egyptians doing the sand dance, in which dancers 'walk like an Egyptian' with their arms and legs in the angular poses of figures in wall paintings and reliefs. The sand dance was, in reality, a comedy dance routine invented in the 1930s.

Ceremonial music

Music had an important role in rituals and ceremonies. Blasts from ceremonial horns would punctuate the religious rites and processions, and noblewomen and priestesses played a type of tinkling rattle called a sistrum at ceremonies in honour of Hathor, the goddess of pleasure, music, dancing, women and love. A horn that was found in Tutankhamun's tomb was sufficiently strong to be blown once more for a famous recording made not long after the tomb was opened. The sound was both high-pitched and eerie in quality. It is thought that music also played a role in the working lives of ordinary Egyptians with, for instance, men tapping out rhythms with sticks as they pressed grapes or performed other routine tasks.

HUNTING

Wealthy Egyptians loved hunting and many hunting scenes were included in the wall paintings of tombs and other sites. The pharaohs themselves were often depicted hunting the hippopotamus-god Seth with a spear in the waters of the River Nile. Hunting hippos was a particularly dangerous sport, as an enraged hippo could easily overturn the flimsy reed boats used by hunters, who might then fall victim to Nile crocodiles. Equally hazardous were the lion and wild bull hunts upon which the pharaohs sometimes engaged in the desert, often pursuing their prey in a chariot and killing it with a spear or bow and arrows. Other game included antelope, oryx, gazelles and hares. Fishing with a spear was also popular with all classes. Fish were also caught using barbed hooks or nets.

Fowl hunts

The Egyptians also hunted water birds in the marshes of the Nile, and especially prized ostriches, as their feathers could be made into elegant fans. Whole families went on duck hunts, bringing the fowl down by hitting them with throwsticks resembling aboriginal boomerangs or otherwise trapping them with nets. They sometimes took cats with them to frighten birds into the air.

The lionhunter

The pharaoh Amenhotep III was credited with killing 102 lions in just 10 years and was also said to have killed 90 wild bulls in a single hunting expedition.

PART SIX

Working life

The prosperity of ancient Egyptian civilization was built on farming and trade and the taxes that could be raised on such activity. The Egyptians became experts at adapting agriculture to the annual flooding of the Nile and made themselves rich through trading the products of their farmers and craftsmen.

Wealth from the land

The work of humble farmers on the fertile banks of the Nile helped pay for the great monuments which were constructed on the orders of the pharaohs.

THE FARMING YEAR

The farming year for the ancient Egyptians had three distinct seasons. The first season (called akhet) began in July with the annual Nile flood. The mud left by the retreating waters was rich enough to grow plants on a strip of land about 10 km (6 miles) wide. There was also fertile land surrounding oases in the desert. Farmers knew that a good flood would be followed by a good harvest; a bad flood could lead to famine.

PLANTING AND IRRIGATION

The second season (called peret) in the farming year involved the setting of seed in the newly refreshed soil. Priests advised farmers on when to do this, based on the position of the moon and stars. The seeds arrived originally with settlers from the Middle East.

Irrigation of the growing crops with water was vital. Farmers irrigated their crops by digging ditches to get water from the river to their fields. They also used a device called a shaduf to raise water from the ditches. This comprised an upright wooden frame supporting a long pole acting as a lever, with a bucket at one end and a counterweight at the other: it is estimated that a single shaduf could raise over 2500 litres (550 gallons) of water per day with relatively little effort.

HARVEST

After several months tending the growing crops, which included barley, beans, dates, grapes, figs and wheat, the farmers prepared for the third season (which was called shemu) in the agricultural calendar: the harvest.

The collected grain was stored in bell-shaped granaries and recorded by scribes so that the pharaoh's officials knew how much tax farmers should pay. Many farmers also raised long-horned cattle and other livestock as well as geese and bees.

When the Nile was in flood

As no work could be done while the river was in flood between the months of July and November, most of the population of Egypt was free to work on building the pharaoh's palaces, pyramids, temples and other major construction projects.

The last Nile flood

The age-old pattern of farming based on the flooding of the Nile lasted until the 1960s, when the building of a huge dam at Aswan controlled the river's flow.

 Overleaf: A wall painting of an Egyptian farmer ploughing his fields with a team of oxen.

CRAFTS

Because Egypt's farmers were so expert at providing food for the population, others were free to develop different skills as craftworkers. The products made by these Egyptian craftsmen ranged from everyday agricultural tools and household furniture to fabulous golden jewellery and elaborately decorated obelisks which were ordered by the pharaoh himself.

TYPES OF CRAFTS

Egyptian craftworkers were respected for their skills and often lived in their own special area of a town or village. They usually produced goods in their own homes, or otherwise they worked in workshops that were attached to temples or palaces. These skilled workers included stonemasons, artists, carpenters, glass makers, jewellers, perfume makers, potters, sculptors, cobblers, basket makers and weavers (who were mostly women).

Artistic styles

Though it is sometimes possible to tell one craftworker's output from another, artistic styles changed remarkably little throughout the course of Egyptian civilization.

Materials

The Egyptians worked in a wide range of materials, including stone, wood, gold, copper, bronze and leather. They also fashioned baskets and mats and other items from papyrus reeds. It seems that the Egyptians learned how to make linen cloth from the fibres of the flax plant as early as 5000 BC.

WORKING IN STONE

Stonemasons were particularly important in the building of the great monuments of ancient Egypt. Objects made of stone extended from beads and storage jars to massive granite columns and statues. Some craftworkers specialized in carving hieroglyphics and pictures (called reliefs) on stone. These reliefs might be raised (the surrounding stone was chipped away) or sunk (cut into the surrounding surface).

POTTERY

The potters of ancient Egypt made a wide variety of pots, jars and other items. Some were intended for domestic use, as containers for food and drink. Others were meant to serve as grave goods. They worked with Nile mud (*iqdon*) mixed with cow dung and straw, and thrown onto a potter's wheel, then decorated and baked (fired) in a kiln. Clay was also used to make bricks, crockery, statuettes, beads and toys.

TRADE

The ancient Egyptians were great merchants and from
an early date traded goods with neighbouring villages
and with peoples from foreign lands. As the Egyptians
did not use money, goods were either exchanged
for other goods or were weighed in balances using
copper weights (called deben) and traded for copper.
Gold and silver could also be used to buy things.

FOREIGN TRADE

Early traders used the Nile to exchange such goods
as clay pots, linen, beer, bread and salt with their
neighbouring villages up and down the river. Trade
links were later established further afield, with the
merchants of ancient Egypt venturing hundreds of
miles into Africa and around the Mediterranean region,
where they took their grain, gold, linen, papyrus, cattle,
salted fish and other goods.

Exotic goods

The Egyptians exchanged their goods for, among other
products, ebony and ivory from Africa, gold from Nubia,
silver from Syria, cedar wood, oils and horses from
Lebanon, copper from Cyprus and the Sinai Desert and
lapis lazuli from Afghanistan. They also brought back
incense and exotic animals from the mysterious Land

Taxation

Egyptian merchants had to pay taxes on the goods that they traded and, in so doing, they helped to make Egypt the richest country in the ancient world. Their activities thus funded the building of many of the great monuments that can still be seen today.

of Punt, which is thought to have been somewhere at the far end of the Red Sea. Other imports included spices, olive oil, leopard skins, elephant tusks, giraffe tail fly whisks, ostrich feathers, apes and slaves.

 Wall painting of the weighing of gold, using a set of scales and a weight in the shape of a bull's head

TRAVEL AND TRANSPORT

Successful trade with foreign lands depended upon the ability of the Egyptians to transport their goods by land and sea. Ancient Egypt lacked a proper road system, as most travel was undertaken by boat up and down and across the crocodile-infested Nile, initially using small boats made from papyrus reeds. Even relatively poor families would own such a boat, which they could use for travelling to market or hunting wild fowl.

BOATBUILDERS

The Egyptians soon became expert boatbuilders, constructing sturdy, wide-bodied wooden boats with distinctive high prows often ending in a round lotus flower shape. Powered by a combination of oars and sails, these boats were capable of sailing long distances and the largest could transport heavy cargoes such as obelisks and stone blocks for the pyramids.

Red Sea canal

Around 490 BC the Egyptians completed the digging of an 85-km (53-mile) canal joining the Nile with the Red Sea. This made it much easier for them to carry their cargoes of goods to markets in east Africa and Arabia.

OVERLAND JOURNEYS

For short overland journeys the early Egyptians relied largely on donkeys to carry their goods, although the animals they kept were not usually big enough to carry people. A later alternative was the camel, which could transport goods across the desert to isolated oases.

Wheeled vehicles were not introduced in Egypt until around 1550 BC, and even then, lacking proper roads, the use of carts was limited by the soft sand of the surrounding deserts, into which wheels easily sank.

 The waters of the Nile provided the easiest means of communication between different parts of Egypt.

PHAROS

PART SEVEN

Learning in ancient Egypt

The ancient Egyptians were notable among the world's earliest societies for their extensive scientific knowledge and their many cultural achievements, which were to influence many later civilizations, including those of Greece and Rome.

The Pharos of Alexandria
Egyptian scholarship made possible such achievements as the Pharos (lighthouse) of Alexandria, which was one of the Seven Wonders of the Ancient World.

HIEROGLYPHICS

The ability to write things down was essential to the emergence of ancient Egyptian civilization. Around 3000 BC the ancient Egyptians developed a writing system called hieroglyphics (although in their own mythology the gift of writing was presented to them by Thoth, the god of wisdom).

In hieroglyphics, each symbol (or hieroglyph) was a stylized picture representing a particular word or sound. These could represent objects, names, ideas or even numbers. There was a total of around 700 hieroglyphs, all of which had to be learned by the scribes (writers). Many of the symbols were based on real objects, such as snakes, hands or birds, and were commonly believed to have magical properties.

EVERYDAY WRITING

Thousands of examples of hieroglyphic writing have survived, not only on the walls of temples and tombs but also in numerous papyrus documents, ranging from letters to official reports. There was more than one type of hieroglyphic writing, and the hieroglyphs used in such everyday documents were generally written (from right to left) in simplified hieratic script. Because this version of hieroglyphics used less

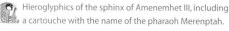

Hieroglyphics of the sphinx of Amenemhet III, including a cartouche with the name of the pharaoh Merenptah.

Cartouches

The names of pharaohs were always contained in an oval-shaped frame called a cartouche. Such cartouches appear frequently in temples and other important sites, sometimes in lists of the ruling pharaohs of Egypt.

elaborate pictures it could be written much more quickly. Another even simpler, and thus more quickly written, form called demotic script was later developed for use in everyday dealings. Much of our knowledge of ancient Egypt is based on translations of these hieratic and demotic scripts.

HIEROGLYPHIC ALPHABET

There follows a selection of ancient Egyptian hieroglyphics roughly equivalent to the letters of the modern alphabet.

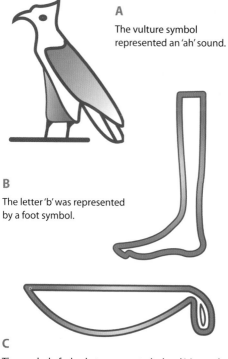

A

The vulture symbol represented an 'ah' sound.

B

The letter 'b' was represented by a foot symbol.

C

The symbol of a basket represented a hard 'c' sound.

D

The letter 'd' was represented by a hand symbol.

E

A single reed represented an 'ee' sound (or 'ah' if placed at the beginning of a word).

F

The letter 'f' was represented by a horned viper.

G

A hard 'g' sound was represented by a jar stand.

H

The letter 'h' was represented by a reed shelter, or by a symbol resembling a length of twisted flax.

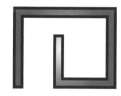

I

The nearest equivalent to a modern 'i' was the symbol of a single reed.

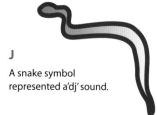

J

A snake symbol represented a 'dj' sound.

K

The sound made by a modern 'k' was represented by a basket symbol.

L

A lion symbol represented the modern 'l'.

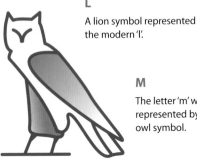

M

The letter 'm' was represented by an owl symbol.

N

The letter 'n' was represented by a wavy line, suggesting rippled water, or else by a crown symbol.

O

The nearest equivalent to a modern 'o' was the image of a quail chick.

P

The letter 'p' was represented by a mat symbol.

Q

A symbol suggesting the slope of a hill represented the sound made by a modern 'q'.

R

The letter 'r' was represented by the image of a mouth, or else by that of a lion.

S

The symbol representing the sound of a modern 's' was based on that of a folded cloth.

T

The semicircular symbol representing a 't' sound was actually that of a loaf.

U

The symbol of a quail chick represented an 'oo' sound.

V

A 'vee' sound was represented by a horned viper symbol followed by a single reed.

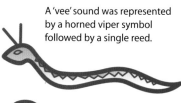

W

The nearest equivalent to a modern 'w' was the symbol of a quail chick.

X

The sound represented by a modern 'x' could be represented by the symbols of a basket and a folded cloth.

Y

The modern 'y' was represented by a pair of reeds.

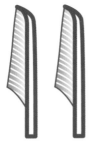

Z

The Egyptian equivalent of a modern 'z' was a door bolt symbol.

PAPYRUS

As paper had not yet been invented, the ancient Egyptians devised their own kind of writing material, called papyrus after the plant it came from (also the origin of the modern word 'paper'). Making papyrus was a complicated and time-consuming process, meaning that the finished product was expensive and it was generally only used for important documents.

PAPER-MAKING

The first stage of the paper-making process involved cutting some of the papyrus reeds that grew plentifully on the banks of the Nile and peeling them. The inner fibres were then cut into thin slices and laid flat in rows on a frame, a second row crosswise over the first, and hammered or pressed until the plant juices stuck the strips together. After being allowed to dry, the surface was then rubbed smooth with a flat stone. Separate pieces of papyrus could be glued into long strips and rolled up in long lengths (called scrolls) or cut into

Poor man's paper

Because papyrus was expensive, everyday messages were more likely to be scribbled on pieces of broken pottery.

pages and then made into books. The papyrus of the ancient Egyptians proved to be extremely long-lasting, and thousands of such documents have survived to the modern day.

Writing materials

Pens to write with were made from reeds with frayed ends. The ink used by the scribes was made either from charcoal, soot, crushed minerals or from red earth which was mixed with water.

 This sample of papyrus sheet is blank and ready to be written upon.

KNOWLEDGE AND EDUCATION

The Egyptians understood the value of education, and they organized schools at their temples for the training of the male children of wealthy families and high officials. Girls of wealthy families were taught at home. Children of poorer parents, who tended to follow the trade of their fathers, received no formal education, and most people could not read or write.

TRAINING OF SCRIBES

Children admitted to the temple schools were strictly disciplined by the priests, who trained them as scribes (writers) from around the age of nine. At the end of their schooling pupils would be able to undertake responsible jobs, such as making records of the size of the yearly harvest and calculating the amount of tax to be paid to the pharaoh. They might also be given important posts within government departments, or responsibility in legal or military matters.

Viziers and senior posts

Scribes occupied positions of respect and could rise to senior posts in the royal court, including that of vizier. They collected taxes, settled legal disputes, designed buildings and recruited men for the army. Perhaps the most famous of all the scribes to emerge from the

Storehouse of knowledge

The Egyptians were celebrated for their great libraries, in which thousands of scrolls on a huge range of subjects were stored. Particularly famous was the library of Alexandria, founded by the Ptolemies, which much influenced early Greek civilization.

Egyptian education system was the vizier Imhotep. Such was his reputation for wisdom and administrative skill that after his death he came to be considered a god.

Although the majority of the scribes were male, the ancient Egyptians also had a word meaning 'female scribe', so some women were presumably allowed to do such work, too.

Statue of the scribe Ka-Irw-Khufu, found in his tomb at Giza in 1957.

SCIENTIFIC KNOWLEDGE

The ancient Egyptians made many significant advances in scientific knowledge in fields that ranged from agriculture and astronomy to medicine and mechanics.

AGRICULTURE

Predicting the coming of the annual Nile flood was crucial to agriculture and a prime preoccupation of Egyptian scholars. In order to analyse water levels, they devised nilometers, which were measuring points on the bank of the river at which accurate measurements could be taken of the depth of water at different times.

Irrigation of the fields was also made much easier through the invention of a device called the shaduf, which enabled the farmers to raise buckets of water from the River Nile to water their crops.

ASTRONOMY

Egyptian scholars made extensive studies of the stars and based their calendar upon the movement of Sirius, the brightest star in the sky. Their skill is evident in the alignment of their temples and other monuments with certain stars so that their light penetrated inner rooms on certain sacred days.

CALENDARS

In addition to the astronomical calendar based on the movement of Sirius, which happened to reappear annually at the time of the Nile flood, Egyptian scholars also created a 365-day farming calendar, which was divided into three seasons of four months each, and a lunar calendar, which dictated when ceremonies should be performed in honour of the moon god Khonsu. The use of such calendars enabled the priests to advise farmers when the Nile might flood and when they should plant their crops.

MEDICINE

The Egyptians were skilled in the treatment of illnesses and injuries, though they also relied to some extent on magic and superstition. They possessed impressive knowledge of childbirth, dentistry, eye disorders and

The Pharos of Alexandria

As well as the temples and pyramids, other notable proofs of technological achievement in ancient Egypt included the Pharos of Alexandria. One of the Seven Wonders of the Ancient World, this was a massive lighthouse that was erected under the Ptolemies to guide ships safely into the harbour at Alexandria.

tumours and evidently understood the importance of the heart (though not, apparently, that of the brain).

Medicinal remedies

Egyptian physicians used garlic, juniper and other plants to make ointments and medicines and had the knowledge necessary to mend broken bones, as was proved by evidence of healed bones in several mummies. Among the many remedies that have survived to posterity is one that recommends eating garlic bulbs to expel tapeworms.

The Pharos of Alexandria, one of the Seven Wonders of the Ancient World, was the tallest building ever built and the light cast by its mirror could be seen 50km (35 miles) away.

PART EIGHT

War in ancient Egypt

The wealth of ancient Egypt made it a target for its neighbours, but it also provided the funds for sustaining armies to protect and expand the country's borders over the centuries.

Egyptian armies
Under various warrior-pharaohs, the Egyptian armies marched southwards into Africa and eastwards into Palestine.

WARFARE

The ancient Egyptians were not an aggressive people and for much of their history they lived in peace with their neighbours. Periodic attacks from other peoples, however, meant they needed an army for defensive purposes, and many warrior-pharaohs defended and even expanded Egypt's borders by leading military campaigns against the Nubians and other enemies.

THE EGYPTIAN ARMY

The first Egyptian army was established to defend the country around 3000 BC. These early soldiers all fought on foot, as horses and chariots were not introduced until around 1600 BC. Over the centuries, these part-time troops repelled various attacks from surrounding countries, aided by the natural defences of the desert and the lack of proper roads. Egyptian troops were

Inspired generalship

Ramesses II was noted for his legendary courage on the battlefield. At the battle of Kadesh he defeated his enemies virtually single-handed after most of his army had fled. To terrify his opponents still further he often took his pet lion into battle with him.

Severed hands

Among the most grisly tasks faced by Egyptian soldiers was that of cutting off the right hands of dead enemy soldiers. The severed hands were then piled up so that the number of the dead could be counted.

transported quickly by boat up and down the Nile to wherever they were needed to prevent invasion.

Later pharaohs of ancient Egypt conducted military campaigns in Nubia, Palestine and Syria. Their armies were properly organized, with a system of ranks descending from the pharaoh at the head of the army all the way down to the lowest officers, who commanded units of 50 men. Other officials who were attached to the army included military scribes, who recorded the progress of campaigns and wrote official documents and messages.

WARRIOR-PHARAOHS

The most famous of the Egyptian pharaohs to go to war included Ramesses II, who was only 10 years old when he first led his troops into battle. Under his leadership, the army was divided into four regiments named after the gods Amun, Re, Phtah and Seth. Each regiment had infantry, archers and chariot troops.

WEAPONS

The first Egyptian soldiers fought with daggers and swords and axes with stone blades, but they later used weapons made of copper and, later still, bronze. Iron was in short supply and was rarely available for making weapons, although some rare examples have survived.

BOWS AND ARROWS

Some soldiers fought as archers with bows and arrows, which were made of reed and originally fitted with flint arrowheads but later with arrowheads of hard ebony wood or bronze. Archers often wore finger guards and wrist protectors to protect themselves when drawing and releasing the bowstring, which was made of twisted sinew or gut.

Improved designs

The bows used by Egypt's Hyksos invaders in the middle of the second century BC were superior to

Throwing sticks

Another weapon used by the Egyptians was the throwing stick, which was similar to an aboriginal boomerang in shape (although it did not return to the thrower).

 A model of Nubian archers, which was found in an excavated tomb in 1894.

those that were used by Egypt's defenders, but the Egyptians quickly learned to copy these improved bows, which had much greater range than their own, and they used them to help expel the invaders several decades later. The best bows had a composite structure with a strip of sinew and a layer of birch bark laid over a core of wood.

SPEARS AND SHIELDS

Foot soldiers also fought with long spears and carried shields that were made of wood or ox hide. In the later history of Egypt most soldiers, including the pharaohs themselves, wore armour which was made from strips or discs of leather.

EGYPTIAN WAR CHARIOTS

When Egypt was largely overrun by nomadic Hyksos invaders 3600 years ago they owed much of their success to their use of horse-drawn chariots, with which the Egyptians were unfamiliar. The Egyptians, however, soon learned to add similar war chariots to their own army and they eventually drove the invaders out.

 Egypt's warrior-pharaohs typically led their troops into battle in horse-drawn chariots.

Tactics

Chariots proved particularly effective against foot soldiers. Charioteers broke up the enemy formations by charging straight through them and then turning back to fall on the scattered foe from the rear.

Egyptian war chariots carried two men, a driver and an archer. These archers were recruited from the rich or noble families who could afford a chariot and horses, which were very expensive items.

Warrior-pharaohs were often depicted in chariots with bow and arrows, leading their army into battle and crushing their enemies beneath their wheels.

CONSTRUCTION

The chariots themselves were made of wood and were designed to be light and fast. The various parts of the chariot were attached with leather and glue, totally without the use of nails. The joints were secured with rawhide and were probably gilded to keep them waterproof. The lack of any suspension meant that the chariots were difficult to handle, and hitting a target with a bow and arrow must have been difficult.

THE LIFE OF A SOLDIER

In the early years of the ancient Egyptian civilization, the country's soldiers were part-time fighters who took up arms when the pharaoh demanded their help to defend Egypt from attack. Once the danger had passed they returned to their civilian lives.

A PROFESSIONAL ARMY

With the creation of a professional army during the New Kingdom period, soldiering became a profession in its own right and the young men who served in the country's forces were properly trained and equipped. Under the personal leadership of the pharaoh, these soldiers might spend long periods on campaign in distant regions, fighting Egypt's enemies in Nubia and Palestine and elsewhere.

When they were not required for fighting, many soldiers often undertook civilian roles at the pharaoh's command. These tasks including the digging of canals and the transporting of stone for new monuments.

Medals

Soldiers who distinguished themselves in battle might be rewarded with prized golden fly medals recognizing their skill in 'stinging' the enemy.

WAR AT SEA

Not all of ancient Egypt's battles were fought on land. Because of their long experience in trading with foreign countries, the Egyptians were capable sailors and from time to time they used their ships to defeat their enemies at sea. Notable sea victories included battles during the reign of Ramesses III with the Sea Peoples, who were fierce raiders from the north-eastern parts of the Mediterranean.

Egyptian warships were adaptable wooden vessels that could be propelled either by a large square sail or, when there was little wind, by a bank of oars.

 Long experience of sailing the Nile made the ancient Egyptians capable sailors.

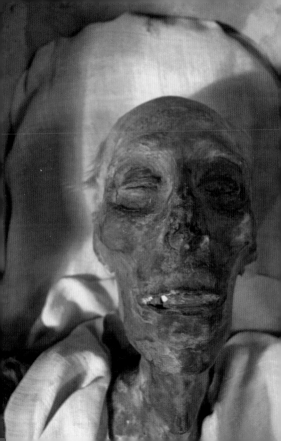

PART NINE

Death in ancient Egypt

The ancient Egyptians believed that when a person died their spirit embarked on a perilous journey to the underworld. It was important that the living did everything they could to ensure this journey went well, by preserving the body in the form of a mummy and burying it in a tomb or pyramid with all the things that the spirit might need in the afterlife.

Life and death
The discovery of ancient Egyptian mummies has contributed much to the understanding of their ideas about life and death.

THE AFTERLIFE

According to the ancient Egyptians, every living person had within their body a life force, which they called their *ka*. The *ka* was created at birth, and when a person died it went into a state of rest, awaiting its reawakening in the afterlife. The *ka* was sometimes depicted in Egyptian funerary paintings as a small figure behind the living person.

Each individual also had a *ba*, which represented their personality. At death it was hoped that the deceased's *ba* would leave the tomb to be reunited with the *ka* and would then be transformed into an *akh* – a fully resurrected spirit.

After death a person's *ba* was believed to journey through an underworld called Duat. In order to protect the spirit of the dead person from such dangers as lakes of fire and deadly snakes while on this journey, spells were written on the coffin and, later, inscribed on papyrus scrolls called Books of the Dead that were buried with the corpse. By reciting these spells the dead person would be kept safe from harm.

The final, and most challenging, test of all came when the dead person's spirit was examined by the gods who ruled in the afterlife.

WEIGHING OF THE HEART

In the Hall of the Two Truths the heart of the dead person was supposedly weighed against his or her past deeds by the jackal-headed god Anubis in the presence of Osiris, the ruler of the underworld, and 42 assessor gods, each representing a district of Egypt.

The verdict of Thoth

The heart was placed on one side of a set of scales and the Feather of Truth on the other, and the verdict was announced by the ibis-headed god of wisdom, Thoth. If the heart was light, the deceased was admitted to the Kingdom of Osiris, thought to be a perfect version of the world above with green fields and plenty of water, to 'become an Osiris' and remain there forever.

The Devourer of the Dead

If the two did not balance, however, then the heart of the dead person would be consumed by a monstrous god who was called the Devourer of the Dead, and the underworld was closed to them.

A place in the stars

Many Egyptians believed that spirits admitted to the kingdom of Osiris made a final journey to the stars, and they were free to roam the earth for eternity.

MUMMIFICATION

The ancient Egyptians believed that the dead would need their physical bodies to house their spirits in the afterlife, and that every care should therefore be taken to preserve their corpses.

The earliest Egyptians wrapped their dead in matting and buried them in the hot sands of the desert, which absorbed water from the remains of the body and prevented further decay. When rich people started burying their dead in stone-lined tombs, however, they found that the bodies rotted in the damp conditions. To solve this, they set about finding out how to dry bodies out completely so that they would be perfectly preserved, a process known as mummification.

PRIESTS OF ANUBIS

The preparation of bodies for burial was undertaken by priests and skilled embalmers, supposedly overseen by Anubis, the jackal-headed god of mummification, who was represented by a priest wearing an Anubis mask. Mummification was a lengthy and expensive

A mummified monkey. Some animals were mummified for religious reasons, but others were family pets thus preserved in order to accompany the dead person in the afterlife.

Mummies

The word 'mummy' itself comes from the Arabic word *mumiya*, meaning 'pitch' or 'bitumen'. It was inspired by the mistaken belief that the black colour of Egyptian mummies was due to the application of such substances.

process, and only pharaohs and the rich could afford to have their mummified remains placed in fine stone tombs or in pyramids. Those who could not afford it hoped that when they died they would be allowed into the sun god Re's golden 'boat of millions', which would carry their spirit (or *ba*) to the underworld.

Mummified animals

Certain sacred animals were sometimes mummified. Archaeologists have found carefully mummified cats, dogs, crocodiles and monkeys, among other creatures.

A profitable business

During the period of the Middle Kingdom, Egypt's increased prosperity meant that many members of the middle class sought to have their bodies mummified. Mummification became a profitable business, though many people probably paid for it as an indication of social standing as much as for any religious reasons.

PREPARING THE DEAD

When a person died, the embalmers had to begin work very quickly indeed, as bodies decayed rapidly in Egypt's hot climate.

EMBALMING PROCESS

The first step in the embalming process was to wash the body in natron, which was a salty solution with antiseptic properties. Next, the embalmers removed the internal organs and placed the liver, the intestines, the stomach and the lungs separately in four special containers (known as canopic jars).

The brain – which the Egyptians thought had no function – was pulled out through the dead person's nose with hooks. The heart, however, was left in place, in order that it could be weighed by Anubis in the afterlife (see page 197).

Mistakes of the embalmers

Embalming was not always successful. When the mummy of Tutankhamun was examined it was found that the priests had put too much ointment on the body, causing parts of it to burn away.

WRAPPING IN BANDAGES

The body was then left to dry out for 40 days. When this period of time had elapsed, it was washed, dried and restuffed with bags of natron, resin and sawdust. The skin was rubbed with ointments, and artificial eyeballs were inserted. The body was then coated in resin and tightly wrapped in linen bandages, beginning with the head, toes and fingers and then progressing to the limbs and torso. Magic charms (called amulets) were placed within the wrappings to protect the deceased. Finally, a face mask resembling the dead person was added.

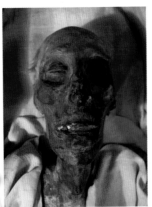

 The skills of the ancient Egyptian embalmers are evident in the well-preserved mummy of pharaoh Ramesses II.

COFFINS

When the wrapping of a mummy was complete it was placed inside a coffin. In the case of rich people, there might be a series of body-shaped coffins, each bigger than the last. They were usually made of plain wood, but in the case of the very wealthy or powerful they might also be fashioned from gilded wood or beaten gold inlaid with lapis lazuli and other gems.

The front and sides of coffins might be completely covered in painted decorations that included various sacred symbols, depictions of the gods and goddesses and spells written in hieroglyphics. The head of the coffin was often adorned with an idealized portrait of the dead person, so that their spirit would identify the body as their own in the afterlife.

STONE SARCOPHAGUS

Once closed, the inner wooden coffins were in turn placed inside a big stone coffin called a sarcophagus, which again might be richly carved with sacred images. Sometimes, as was the case with Tutankhamun, the sarcophagus itself was encased in a series of shrines, which were similarly decorated with sacred symbols and spells and sealed after the body was placed inside them.

FUNERAL RITES

The ancient Egyptians developed detailed funeral rituals that had to be correctly observed if a person's spirit was to complete the journey to the afterlife.

The rites took place 70 days after death (allowing time for the preparation of the mummy) and were attended by grieving relatives and, in the case of wealthy families, groups of hired mourners. Priests sprinkled sacred liquids, chanted incantations and burned incense.

 Careful observance of funeral rites was thought to be essential for a person's welfare in the afterlife.

The pharaoh's boat

The boat that was used to transport a dead king across the Nile was sometimes buried with the pharaoh himself. The funerary boat of King Khufu was found in a pit alongside the Great Pyramid at Giza. Built of cedar wood, it was over 43 metres in length.

FUNERAL PROCESSIONS

The funeral included a procession in which the coffin was transported to the Nile on a bier (a canopied sled pulled by oxen) and then taken across the river by boat (painted green to symbolize nature and rebirth) to be buried on the west bank. Mourners tore their clothes and tossed dust onto their heads to show their grief.

OPENING THE MOUTH

At the tomb the coffin was placed upright for an important ceremony which was called 'opening the mouth'. This involved the dead person's son, or a priest, symbolically opening the eyes, ears and mouth of the body with a special instrument.

This ritual ensured that the dead would be able to speak, see, hear, eat, drink and move around in the afterlife. After the funeral, the tomb was sealed.

TOMBS

Most Egyptians lived on the east bank of the Nile, and the west bank, where the sun set, was traditionally considered the place of the dead. The deceased were said to go to meet the sun when they died, making a symbolic crossing of the Nile, which lay between the lands of the living and the dead. It was on the west bank, therefore, that most Egyptians were buried.

TYPES OF TOMBS

The first Egyptians buried their dead in pits in the desert, where the dry sand would preserve the bodies. Later, wealthier families put their dead in long low tombs that were made of small stones and bricks, called *mastabas*. Next came the era of pyramid-building, in which pharaohs and other important officials were laid to rest in huge stone pyramids, mostly in the region of the capital of Memphis, near modern Cairo.

VALLEY OF THE KINGS

From 2150 BC the pharaohs ceased to be buried in pyramids and were laid to rest in tombs in the Valley of the Kings, situated in the desert between Thebes and Abydos (opposite modern Luxor). Burials continued here for some 500 years. The entrances to most tombs

Guardian of the Valley of the Kings

Tomb-robbers in the Valley of the Kings had to be brave.
As well as human guards, they risked offending the valley's
protector-god, Meretseger, in the form of a deadly cobra.

were hidden and guards were posted to deter tomb-
robbers, although nearly all the tombs were eventually
discovered and plundered. Another very important
burial site was the Valley of the Queens, which is
located in a neighbouring valley.

 The richly decorated tomb of Ramesses VI in the Valley
of the Kings near Luxor.

PYRAMIDS

The most famous ancient Egyptian monuments still standing today are the pyramids built as tombs for dead pharaohs on the west bank of the Nile. The pyramids were the first buildings made by man using large blocks of stone precisely cut for the purpose.

The very first pyramid was that built for King Djoser by his architect Imhotep at Saqqara around 2650 BC. With six levels rising to a flat top, it is usually called the Step Pyramid, the idea being that the spirit of the dead pharaoh would ascend the pyramid like a staircase to meet the sun. It was probably designed, like all pyramids, as a recreation of the mound on which the sun-god stood when creating the universe.

TRUE PYRAMIDS

Later (4500 years ago) came the true pyramids, which had four smooth sides meeting at a central point. Many of these pyramids were faced with gleaming white limestone, which reflected the sun and made a dazzling sight. Most pyramids, including the Bent Pyramid and Red Pyramid of Sneferu at Dahshur and those of Khufu, Khafra and Menkaura at Giza, were built during Dynasty IV, when the rule of the pharaohs was at its strongest.

Counting the pyramids

The precise number of pyramids in Egypt is uncertain, as many have been reduced to rubble. By 2002 archaeologists claimed to have discovered as many as 110 pyramids.

As the power of the pharaohs declined it became more difficult to muster the resources and later pyramids tended to be less well-constructed. The last pyramids were built by 2150 BC, after which later pharaohs were buried in the Valley of the Kings and elsewhere.

THE PYRAMIDS OF GIZA

The most famous of the Egyptian pyramids are the group on the plateau at Giza, outside Cairo. These include the massive pyramids of the pharaohs Khufu, Khafra and Menkaura as well as several smaller ones and various *mastabas* (mud-brick tombs) in which the family, friends and servants of the pharaohs were buried. The Giza complex also includes the ruins of mortuary temples, where offerings were made in honour of the dead kings.

The largest of the Giza pyramids is the Great Pyramid, which was built for King Khufu around 2528 BC. One of the Seven Wonders of the World, it was constructed with 2.3 million limestone blocks, each weighing more than two tons. At 138 m (450 feet) high, the pyramid was originally covered with white limestone casing, but most of this was removed in medieval times for building work in Cairo. Alongside it are three smaller pyramids that were built for the pharaoh's chief wives.

Guardian of the pyramids

Close by the Giza pyramids stands the mysterious Sphinx. This is a huge statue of a lion with the face of a pharaoh, supposedly modelled on that of King Khafra. The origins of the sphinx remain obscure. Over the centuries it was frequently covered in sand and eventually dug out.

BUILDING A PYRAMID

The ancient Egyptians lacked modern tools and to build their pyramids had to rely instead on ingenuity and powers of organization. Tools were simple, but the pharaohs could call upon thousands of workers, who often considered such work a religious duty.

The riddle of the sphinx

In Greek legend the sphinx killed any man who could not solve the riddle she set, but to the ancient Egyptians the sphinx was a symbol of the divine power of a king (the lion being associated with the sun-god Re).

Building materials

The first task was to find suitable building materials and transport them to the building site. The Egyptians quarried limestone, sandstone and granite and mined gold in the desert. They split stones into even blocks by hammering wooden wedges into rows of holes made with copper chisels.

When water was poured onto the wedges they swelled and split the rock, a process that took around 10 hours. The stones were finished using relatively soft copper chisels and saws, as there were no iron tools.

The massive stones were then taken across the Nile on barges and, because wheels were still to be introduced to ancient Egypt, they were carried over land on sleds which were pulled by large teams of men.

Ramps of sand

To check that the base for a pyramid was perfectly flat, the Egyptians (who had no spirit levels) dug narrow channels around the base area and filled them with water. They then levelled the ground until the depth of water was the same in all the channels. The pyramids themselves were probably constructed by means of a huge ramp of sand or earth up which the stone blocks were dragged on rollers by teams of workers and then levered into position. When one layer was finished the ramp was extended to begin the next.

INSIDE A PYRAMID

Most of the interior of a pyramid was filled with stones and rubble. Deep within the structure, however, there were also passages leading to one or more burial chambers, in which the sarcophagus of a dead pharaoh was placed. There might also be false passages and empty chambers designed to mislead tomb-robbers.

Chambers and galleries

Some of the spaces existing within the pyramids were impressively large. Inside the Great Pyramid of Giza, for instance, there is an underground burial chamber (never used after the pharaoh objected to being buried beneath ground level), another abandoned burial chamber and the King's Chamber, in which the sarcophagus of King Khufu was actually placed.

The King's Chamber was reached by a Grand Gallery, which had a stone roof over 8 metres (28 feet) high and a total length of 47 metres (154 feet). After the burial this passage was sealed off with large blocks of granite.

A lifetime's work

The Great Pyramid of Giza took 20 years to build, with around 100,000 blocks being set in place each year (a rate of 285 blocks a day).

It is thought that the pyramids were recreations of the mound on which the sun-god stood to create the universe.

The last workers left by means of an escape passage, which was then closed with stones that were identical to all the others that had been used in the construction so that no one would be able to find the entrance.

A well-kept secret

The entrance to the Great Pyramid remained secret for 3000 years. It was finally located in the ninth century by a local Arab ruler enticed by tales of magic metals to be found within.

LAYOUT OF THE GREAT PYRAMID

For centuries no one knew how to get into the largest of the pyramids, that of King Khufu at Giza. When the location of the entrance was discovered, the pyramid was found to contain a series of chambers which were connected by passageways and shafts.

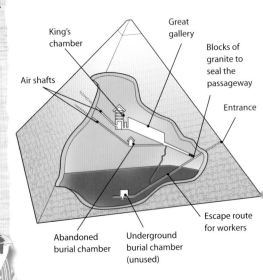

King's chamber

Great gallery

Blocks of granite to seal the passageway

Air shafts

Entrance

Abandoned burial chamber

Underground burial chamber (unused)

Escape route for workers

BURIAL CHAMBERS

Burial chambers within pyramids and other tombs varied from the highly elaborate to the relatively plain in terms of design and decoration. This could reflect the status and wealth of the person who was buried there, but it could also be the result of lack of time for preparation of the tomb, perhaps due to the early or unexpected death of the person concerned.

What they can tell us

The study of ancient burial chambers has contributed much to our understanding of ancient Egyptian culture, especially as a result of the many hieroglyphics, reliefs and paintings with which their walls were often adorned. These have revealed many telling details of everyday life in ancient Egypt as well as providing us with information about religious belief and practices.

Wall paintings

Some tombs included colourful paintings depicting important events in the life of the dead person as well

The dead restored

In tomb paintings of the dead, the deceased person is usually depicted as young and healthy, portraying them as they hoped to be once again in the afterlife.

as members of their immediate family. From these pictures, archaeologists have identified the tombs of a wide range of ancient Egyptians, ranging from viziers and pharaohs to scribes and relatively humble priests and landowners.

GRAVE GOODS

From an early date the Egyptians buried their dead together with various household objects, such as pots and other utensils, which they believed would be useful to them in the afterlife. Some items were just ordinary everyday tools, while others were specially made by craftworkers to be placed in the tombs. Indeed, the pharaohs employed many craftworkers on a permanent basis to fashion precious items to be put into royal tombs.

Types of goods

As well as fabulous jewellery and ornaments, grave goods could include beds, chariots, statues, dried flowers, clothing, oil, cosmetics, musical instruments, models of the sacred boat in which the spirit may sail to the underworld and, in the case of the pharaohs, thrones and crowns.

Canopic jars

The contents of burial chambers also included the four canopic jars containing the internal organs removed

Burial myth

The colourful idea that the servants of a pharaoh were entombed with him when he died in order to serve him in the afterlife is a later invention.

from the corpse during the mummification process. Housed in a small shrine, these were made of pottery with tops variously fashioned in the shape of a human, baboon, dog or falcon head. These organs, it was believed, would be reunited with the rest of the body in the underworld.

MODEL SERVANTS

Also thought essential for the welfare of the deceased was the inclusion of model servant figures, known as *shabtis*. These were expected to come alive to serve the dead person in the afterlife and to do any manual work that Osiris might require of the dead person's spirit.

Some people were buried with hundreds of *shabti* figures, which often held agricultural implements and other tools, as though ready to work. By the Third Intermediate Period many mummies were being buried with a *shabti* for every day of the year, together with overseer figures holding whips, whose job it would be to keep all the other servants under control.

PART TEN

Egypt today

The dry climate of Egypt has meant that many objects thousands of years old have survived more or less intact to modern times. These range from fabulous treasures and other items unearthed by archaeologists over the last 125 years or so to the pyramids, temples and statues that today attract countless visitors to the country.

Tutankhamun
The priceless relics that were found in the tomb of Tutankhamun in the 1920s intensified the modern fascination with ancient Egypt.

DISCOVERING EGYPT

The ancient Egyptian civilization had a profound influence upon the Greeks and Romans, and curiosity about the land of the pharaohs has continued down through the centuries. William Shakespeare and George Bernard Shaw rank among the great minds who have brought ancient Egypt to life in their writings, and the stories surrounding such names as Cleopatra, Ramesses and Ptolemy are widely familiar, inspiring a plethora of plays, novels, poems, paintings, operas and films.

Our knowledge of ancient Egypt is based on a variety of sources. Some details were recorded by classical writers, such as the Greek historian Herodotus (who visited Egypt around 450 BC and described how the Egyptians mummified their dead). Other details, however, remained obscure until relatively recent times when hieroglyphics were eventually decoded.

THE ROSETTA STONE

The key to unlocking the secrets of ancient Egypt lay in translating the hieroglyphics found at sites throughout the country, but for many centuries their meaning was entirely unknown. In 1799, however, the chance discovery by a French soldier of the so-called Rosetta

The first Egyptologist

The man responsible for decoding the Rosetta Stone was the French archaeologist Jean François Champollion. The work took him several years and the effort involved in this and in early excavations in the Valley of the Kings contributed to his premature death at the age of 41.

Stone opened the door to greater understanding. The Rosetta Stone was a large slab of basalt on which was written an account of the coronation of Ptolemy V, inscribed initially in hieroglyphics, then repeated in a simplified form of hieroglyphs known as demotic script and then repeated once more in Greek, which, crucially, classical scholars understood.

Deciphering the hieroglyphics on the Rosetta Stone, now in London's British Museum, opened the door to greater knowledge of ancient Egyptian history and culture.

TOMB-ROBBERS

Making sense of ancient Egyptian history from the objects that were found in pyramids and tombs was complicated at an early date by the activities of the tomb-robbers, who stole anything of value from these sites. Defying a complex system of concealed entrances, false corridors, blocked doorways and armed guards, tomb-robbers had robbed virtually all the pyramids and tombs in the Valley of the Kings by the year 1000 BC. Only the tomb of Tutankhamun remained substantially intact, although even that was raided at least twice and had to be resealed.

WHO WERE THEY?

Tomb-robbers were often the same men who had been employed to construct and fill the tombs in the first place. Such was their knowledge of tombs that there was even a book, called *The Book of Buried Pearls*, which offered advice on getting past the spirits that were believed to guard the dead.

DAMAGE

The activities of the tomb-robbers were very destructive. Having burrowed their way into an underground tomb or pyramid, they wrenched the lid off the sarcophagus

Impaled on a stake

Tomb-robbers in ancient Egypt risked a terrible death if they were caught. They would have the soles of their feet beaten and would then be put to death by being impaled on a sharpened wooden stake.

and then broke open the coffins within. When they reached the mummy itself, they hacked off the gold death mask and tore at the wrappings in search of valuable amulets and other pieces of jewellery. Any gilding on the coffin was prized off and removed, together with anything else the thieves could carry. In some cases they set fire to the burial chamber and later returned to retrieve the puddles of melted gold from the ashes.

STOLEN MUMMIES

Some tomb-robbers were so greedy that they even stole the mummies themselves from their tombs. Some of these mummies were later sold on to early tourists as souvenirs. A mummy bought by an American visitor in the mid-nineteenth century and displayed as a curiosity for many years in a museum at Niagara Falls was eventually identified through X-rays and computer analysis as that of Ramesses I and returned to Egypt as a gesture of goodwill in 2003.

EGYPTIAN ARCHAEOLOGY

Scholars began to show interest in unearthing what might remain of ancient Egyptian civilization early in the nineteenth century, when many Europeans arrived in the country while fighting in the Napoleonic Wars.

Academic and general interest intensified following the decoding of hieroglyphics in the 1820s , which greatly increased our knowledge of how the ancient Egyptians lived, and Egypt later became a popular tourist destination which was visited by professional and amateur historians and archaeologists alike. Serious archaeological exploration of the country, however, dates from the end of the nineteenth century.

FINDS

Since the nineteenth century, archaeologists have explored scores of pyramids, tombs and half-buried temples, recovering thousands of important objects, including mummies, statues, jewellery and even wooden figures and fragile papyrus documents.

Preservation

Although few buildings and other items that were buried in the earth near the Nile have survived because of repeated flooding, objects that were covered by the

Spoils of war

The French emperor Napoleon himself was among those people in the nineteenth century who were fascinated by the pyramids and other finds. On his orders, many items were brought back to France by his conquering armies.

hot sand of the desert are often found to be perfectly preserved, entire temples having been protected for many centuries under banks of drifting sand.

Noted Egyptologists

Important names in the history of Egyptology have included Jean François Champollion, the first translator of Egyptian hieroglyphics; the French archaeologists Auguste Edouard Mariette, sometimes called the 'father of Egyptology', and Gaston Maspero; and the British archaeologists William Flinders Petrie, who explored the Great Pyramid at Giza, and Howard Carter, finder of Tutankhamun's tomb.

DIAGNOSTIC TOOLS

Modern archaeologists have a wide range of advanced technological tools at their disposal. These include X-ray technology, which enables experts to examine mummies internally without damaging them, and also ground-penetrating radar.

NOTABLE DISCOVERIES

The distinct threat that the tomb-robbers and amateur archaeologists of the mid-nineteenth century might unearth many of Egypt's most precious relics without any official supervision led to attempts to control such activity by the Egyptian authorities. This became particularly important after 1881, when it was discovered that tomb-robbers had been plundering an extraordinary tomb containing dozens of mummies.

DEIR EL-BAHRI

The tomb at Deir el-Bahri, in a neighbouring valley to the Valley of the Kings, was uniquely important. As well as some 6000 smaller items and the mummies of senior priests and other individuals, it contained around 40 royal mummies of the New Kingdom. Apparently rescued from the Valley of the Kings around

Royal tombs

By 1922, archaeologists had found around 60 tombs and burial pits in the Valley of the Kings, although these finds were all eclipsed by Carter's discovery that year of the valley's one largely untouched tomb, that of Tutankhamun.

930 BC during Dynasty XXII and reburied in an attempt to keep them safe from tomb-robbers, the mummies originally included those of the pharaohs Ramesses I, Seti I and Ramesses II among others, although some were now missing.

When proper archaeological exploration of the Valley of the Kings began in 1898 only one archaeologist was permitted to work in the valley at a time. The first permit-holder, the wealthy US archaeologist Theodore Davis, found 35 tombs before giving up his permit in 1915, thinking that the valley was exhausted. The permit then passed to Englishman Howard Carter.

Important finds since Carter's time have included buildings, statues and hundreds more mummies, including one (found in 2003) dating back 5000 years.

Howard Carter at work in the Valley of the Kings.

VALLEY OF THE KINGS

The Valley of the Kings, which is situated on the west bank of the Nile opposite Thebes and Luxor, was used as a royal cemetery for 500 years. It included the tombs

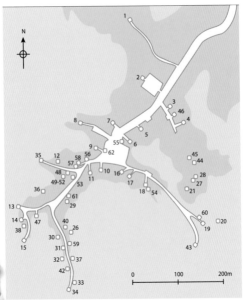

of many famous pharaohs, including that of the boy king Tutankhamun. Although it had been declared exhausted by archaeologists as early as 1912, tombs continue to be found in and around the valley.

The Valley of the Kings

(key for map opposite)

1 Ramesses VII
2 Ramesses IV
3 Son of Ramesses III
4 Ramesses XI
5 Sons of Ramesses II
6 Ramesses IX
7 Ramesses II
8 Merenptah
9 Ramesses V/VI
10 Amenmesse
11 Ramesses III
13 Bey
14 Twosret/ Sethnakht
15 Seti II
16 Ramesses I
17 Seti I

18 Ramesses X
19 Mentuherkepshef
20 Hatshepsut
34 Thutmose III
35 Amenhotep II
36 Malherpri
38 Thutmose I
42 Thutmose II or wife of Thutmose III
43 Thutmose IV
45 Userhet
46 Yuya and Thuya
47 Siptah
57 Horemheb
62 Tutankhamun

Overleaf: The Valley of the Kings, with the tomb of Tutankhamun in the foreground.

THE TOMB OF TUTANKHAMUN

The greatest archaeological find of all was that of the tomb of Tutankhamun, which was located in the Valley of the Kings, by the British archaeologist Howard Carter in 1922. Carter had worked in the valley for several years without finding very much. He was convinced, however, that somewhere nearby must be the tomb of the boy-pharaoh Tutankhamun and he persuaded his patron Lord Carnarvon to fund one last season of digging in the valley.

WONDERFUL THINGS

On 4 November 1922, a young water-carrier for Carter's workers uncovered a stone step, which was the start of a sunken stairway leading to a sealed door bearing the seal of Tutankhamun. Carter summoned Carnarvon from Scotland and on 25 November the door was finally opened, to reveal a corridor filled with limestone chippings. Workmen removed the chippings and the archaeologists reached a second sealed doorway.

The tomb had clearly been broken into by tomb-robbers in the distant past, but when Carter looked through a hole made in the second doorway he saw the glint of gold. When Carnarvon asked if he could see anything, he replied, 'Yes, wonderful things!'

The four chambers of the tomb were filled with many priceless treasures, all of which had to be carefully recorded, numbered and preserved before being displayed in the Cairo Museum. In all it took Carter 10 years to empty the tomb. One of the gilded coffin cases was later replaced in its sarcophagus in the tomb, beneath a protective plate glass lid.

The layout of the tomb

One of the reasons why the location of Tutankhamun's tomb remained unknown for some 3000 years was its position immediately below that of the pharaoh Ramesses VI, which was cut into the rock above it around 200 years later.

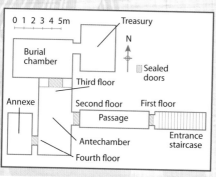

TREASURES OF TUTANKHAMUN

The tomb-robbers who broke into the tomb of the boy pharaoh Tutankhamun stole only small items that they could easily carry and left the rest behind.

The first room explored by the archaeologists was the antechamber, which contained a jumble of over 2000 objects. As well as caskets, boxes and alabaster vases, it included three gilded couches, chariots (stored in pieces to save space) and bouquets of flowers.

GOLD DEATH MASK

Between two black life-sized statues was a sealed doorway leading to the burial chamber. The pharaoh's mummy, complete with a solid gold death mask, had been placed inside a coffin of beaten gold, which had in turn been enclosed within two larger gilded coffins and then a stone sarcophagus. The sarcophagus itself

The curse of King Tut

The death of Lord Carnarvon on 4 April 1923, four months after the discovery of the tomb, led to rumours of a curse associated with it. Carnarvon actually died from pneumonia after a mosquito bite on his face became infected, and the curse was an invention of journalists.

 The marvellous solid gold death mask of the boy pharaoh Tutankhamun, which was found in 1922.

was housed in a series of four shrines, each one bigger than the last, filling virtually the entire room.

THE TREASURY

Beyond the burial chamber was the treasury, which contained another 500 objects, including figures of Tutankhamun, the canopic shrine containing his mummified internal organs, a statue of Anubis, model boats and numerous *shabti* servant figures to serve the pharaoh in the afterlife.

THE ANNEXE

The last room to be cleared was the annexe, which was positioned off the antechamber. This was heaped with a veritable jumble of household items, such as beds, stools, a fan, a sandal, a glove, a razor, ointments, some oil, food and wine, as well as statuettes, more *shabtis*, various weapons, musical instruments and a senet board game.

Personal treasures

Among the most moving finds were two mummified foetuses (probably the stillborn babies of Tutankhamun and his wife Ankhesenamun) and a lock of hair from Tutankhamun's grandmother Queen Tiye.

 The packed interior of the tomb of Tutankhamun, as discovered by Howard Carter and his team in 1922.

ARCHAEOLOGICAL SITES

Many of the most impressive monuments of ancient Egypt can still be seen today, and archaeological tourism is an important part of the Egyptian economy.

GIZA AND SAQQARA

The most notable sites include the Pyramids of Giza which are situated near modern Cairo, and rank high among the wonders of the ancient world. Although most of the limestone casing has now gone, they still attract thousands of tourists every year. Visitors can go inside the pyramids, but they are no longer allowed to climb to the top, as they once were.

Associated with the pyramids themselves are the nearby Sphinx, which is still imposing despite being once used as a target for artillery practice, and the funerary boat of King Khufu, which was found in a pit at the foot of the pharaoh's pyramid.

River of time

Perhaps the most significant relic of ancient Egypt is the River Nile itself, which is still as central to life in modern Egypt today as it was 5000 years ago.

The oldest of the pyramids, the Step Pyramid of King Djoser at Saqqara, has also survived to modern times in relatively good condition, together with its various mortuary temple complexes.

VALLEY OF THE KINGS

Much further to the south near Luxor, the Valley of the Kings and the surrounding area continues to be excavated by archaeologists, and many new tombs have been found in recent years.

Visitors can view the sarcophagus of Tutankhamun in his famous tomb, but most of the other tombs that are open to the public are now empty. Many of the precious artefacts and mummies from these tombs can be viewed in the archaeological museum in Cairo or in other collections elsewhere in the world.

MEMPHIS AND THEBES

Other sites that are much visited by tourists include the ruins of Memphis (the site of a colossal fallen statue of Ramesses II and an alabaster sphinx) and various locations in the vicinity of ancient Thebes (modern Luxor). Alexandria, by contrast, has relatively few archaeological remains, and nothing can be seen of the Pharos lighthouse, although the Greco-Roman Museum in the city contains many pharaonic relics.

TEMPLES OF ANCIENT EGYPT

Some of the temples that were built by the ancient Egyptians have survived in remarkably good condition, having been preserved under the hot drifting desert sand for many centuries.

WELL-PRESERVED RUINS

The most complete ruins to be seen today include the Temples of Osiris and Ramesses II at Abydos, the Temple of Hathor at Dendera, the Great Temple at Luxor, the Temple of Hatshepsut in the Valley of the Queens, the temple of Khnum at Esna, the temple of Horus at Edfu (the best preserved of them all) and the Temples of Philae (reassembled on the island of Agilkia).

Great Temple of Amun

Perhaps the finest of all the ruined temples, however, that you can still visit today is the Great Temple of Amun at Karnak. Continually altered and enlarged by different Egyptian pharaohs over a period of about 2000 years, its many fascinating features include six massive pylons (entrance buildings), sphinx-lined avenues, a sacred lake and the famous hypostyle hall, with 134 closely-packed gigantic columns. Modern visitors to the temple complex at Karnak can enjoy a stunning sound and light show.

Abu Simbel rebuilt

During the period from 1964 to 1968, the entire complex of Abu Simbel was painstakingly moved, piece by piece, by UNESCO and was then reassembled 60m (200ft) higher up the cliffs to escape the rising waters of a new reservoir.

ABU SIMBEL

Unique among Egypt's ancient temples are the two temples of Ramesses II carved into sandstone cliffs overlooking the Nile at Abu Simbel. The Great Temple was built by Ramesses to celebrate the thirtieth year of his reign and dedicated to the sun-god Re (although the four huge statues that guard the entrances are of Ramesses himself, and the surface is covered with hieroglyphics recounting his military feats). Entrances lead to three halls containing more statues and victory scenes. The Great Temple was sited so that on just two days of the year the rays of the sun came through the entrance and lit the inner sanctuary.

The Small Temple of Abu Simbel was dedicated by Ramesses to his wife Nefertari and is guarded by four huge statues of Ramesses and two of Nefertari.

 Overleaf: The amazing temple of Ramesses II at Abu Simbel with its huge statues carved into the rock.

STATUES AND OBELISKS

Still to be seen at many ancient sites are numerous statues depicting the pharaohs and the gods they worshipped, some of them very large. Among the most famous are the huge, toppled limestone statue of Ramesses II at Memphis, four more statues of the same pharaoh in the funerary temple which is known as the Ramesseum near the Valley of the Kings, and those of Ramesses II at Abu Simbel – around 18m (60ft) in height. Elsewhere, visitors may admire a fine statue of the falcon-god Horus at Edfu and two enormous seated statues in a farmer's field at Thebes. These two 18m (60ft) high sandstone giants represent the pharaoh Amenhotep III and were once part of a temple long since destroyed by earthquakes.

London's obelisk

The Egyptian obelisk on the Victoria Embankment beside the Thames in London is popularly known as Cleopatra's Needle. A gift to Great Britain in the nineteenth century, it was actually made for Thutmose III and was nearly lost at sea while being transported on a seagoing barge. It is one of a pair, its twin now standing in Central Park, New York.

OBELISKS

The ancient Egyptians were also famous for erecting massive stone columns called obelisks, which they covered with hieroglyphics honouring gods and kings. The Great Temple of Amun at Karnak has fine examples of obelisks erected by Queen Hatshepsut and others. Two enormous obelisks also once stood in the temple at Luxor, although there is now only one. The other was presented to France in 1831 and stands in the Place de la Concorde in the centre of Paris.

One of the obelisks at Karnak outside the Great Temple of Amun.

MUSEUMS

Many priceless treasures that were found in tombs and at other ancient Egyptian sites are now preserved for posterity in museums around the world.

EGYPTIAN MUSEUM IN CAIRO

The greatest collection of all is the one kept at the Egyptian Museum in Cairo, which was built specially in the mid-nineteenth century to house the many ancient objects that were being found by the first generation of Egyptologists.

Treasures

The museum's vast collections include many of the mummies that were recovered by archaeologists from pyramids and tombs throughout Egypt together with a variety of grave goods, sarcophagi, statues and countless other items.

Tutankhamun on tour

When the treasures found in the tomb of Tutankhamun were toured to Britain and other parts of the world in the 1960s and 1970s, millions of people flocked to marvel at their historical and artistic worth.

Online museums

Most of the major museums that possess collections of objects from ancient Egypt have websites offering pictures and further information about items on display.

Most prized of all are the priceless treasures that were discovered in the tomb of Tutankhamun in 1922, among them the young pharaoh's famous solid gold death mask. Other remarkable exhibits include statues of the pharaoh Akhenaten, the coffin of Ramesses II and numerous well-preserved tools and other objects that were in daily use up to 5000 years ago.

OTHER MUSEUMS

Further ancient relics are housed in the Greco-Roman Museum in Alexandria, the museum at Luxor and other smaller local museums. Outside Egypt itself, there are important collections of ancient Egyptian antiquities in the British Museum in London and elsewhere. The celebrated painted bust of Queen Nefertiti, for instance, is kept in Berlin's Egyptian Museum.

The British Museum's most prized Egyptian artefacts include the Rosetta Stone, which proved the key to unlocking the secret of reading hieroglyphics, and a substantial collection of mummies.

FIND OUT MORE

BOOKS

Amazing Facts about Ancient Egypt, James Putnam, 1994

Ancient Egypt, Barry J. Kemp, 2005

Ancient Egyptian Designs, E. Wilson, 1991

Ancient Egypt Resource Book, James Mason and Sallie Purkis, 1991

Awful Egyptians, Terry Deary, 2006

Egyptian Food and Drink, H. Wilson, 1988

Egyptian Textiles, Rosalind Hall, 1990

Eyewitness Guides: Ancient Egypt, George Hart, 1990

Eyewitness Guide: Mummy, James Putnam, 1993

Great Buildings: The Great Pyramid, Hazel Mary Martell, 1997

How Would You Survive as an Ancient Egyptian?, Jacqueline Morley, 1993

I Wonder Why Pyramids Were Built, Philip Steele, 1995

Look Inside an Egyptian Tomb, Brian Moses, 1997

Making History: Egypt in the Time of Rameses II, Jacqueline Morley, 1993

100 Things You Should Know About Ancient Egypt, Jane Walker, 2001

Technology in the Time of Ancient Egypt, Judith Crosher, 1997

The Complete Gods and Goddesses of Ancient Egypt, Richard H. Wilkinson, 2003

The Complete Pyramids, Mark Lehner, 1997

The Complete Royal Families of Ancient Egypt, Aidan Dodson and Dyan Hilton, 2004

The Complete Temples of Ancient Egypt, Richard H. Wilkinson, 2000

The Complete Tutankhamun, Nicholas Reeves, 1995

The Complete Valley of the Kings, Tombs and Treasures of Egypt's Greatest Pharaohs, Nicholas Reeves and Richard H. Wilkinson, 1996

The Egyptians, Rosemary Rees, 1994

The Myths of Isis and Osiris, J. Cashford, 1993

The Oxford History of Ancient Egypt, Ian Shaw, 2003

The Penguin Historical Atlas of Ancient Egypt, Bill Manley, 1996

The Usborne Encyclopedia of Ancient Egypt, 2004

Tutankhamun: The Life and Death of a Pharaoh, David Murdoch, 1998

WEBSITES

http://archaeology.about.com/od/ancientegypt/
(information about Egyptian archaeology)

http://www.ancientegypt.co.uk
(general information provided by the British Museum)

http://www.ancient-egypt.org
(general information)

http://www.bbc.co.uk/history/ancient/egyptians/
(general information)

http://www.crystalinks.com/cd.html
(information about recent archaeological finds in Egypt)

http://www.newton.cam.ac.uk/egypt/
(guide to further online resources)

http://www.si.umich.edu/CHICO/mummy/
(information about ancient Egyptian mummies)

http://www.touregypt.net/kings.htm
(information about ancient Egyptian rulers)

GLOSSARY

Akhet The season in which the Nile flooded, roughly August to October.

Amulet A piece of jewellery worn in the belief that it would keep evil spirits away.

Ankh Sacred symbol in the shape of a cross with a looped head, representing life.

Ba In ancient Egyptian belief, the mind-spirit of a dead person.

Canopic jar Stone jar in which the internal organs of a dead person were stored after death.

Cartouche An oval frame surrounding the name of a pharaoh written in hieroglyphics.

Chariot A fast, two-wheeled, horse-drawn vehicle, used by the ancient Egyptians in hunting and fighting.

Death mask A mask placed over the face of a mummified body, often decorated to resemble the dead person.

Demotic script A simplified form of hieroglyphic writing developed in the later history of ancient Egypt.

Djed Sacred symbol representing the god Osiris.

Dynasty A ruling house, usually all members of the same family.

Egyptologist An expert in the history of ancient Egypt.

Girdle of Isis A sacred symbol representing the goddess Isis.

Hieratic script A simplified form of hieroglyphic writing predating demotic script.

Hieroglyphics A form of pictorial writing used by the ancient Egyptians.

High priest The senior priest in a temple.

Holy of holies The most sacred shrine in a temple, usually dedicated to a particular god or goddess.

Hyksos A people who invaded Egypt from Palestine

at the time of the Dynasties XV and XVI.

Hypostyle hall A large hall in a temple with a roof supported by many large stone pillars.

Ibis A marshland bird, common on the Nile, that was considered sacred by the ancient Egyptians.

Ka In ancient Egyptian belief, the body-spirit of a dead person, which required a physical body to survive in the afterlife.

Kemet The name, meaning 'black land', given by ancient Egyptians to the fertile land on which they grew their crops.

Kohl A type of black eye make-up made from iron ore.

Linen A type of cloth made from flax.

Lotus Flowering marshland plant adopted as the symbol of Upper Egypt.

Lower Egypt The northern part of Egypt, including the Nile Delta region.

Mastaba A tomb of mud bricks, often dating back to

the early history of ancient Egypt.

Middle Kingdom Period of ancient Egyptian history roughly covering the years 2040 BC–1640 BC.

Mortuary temple A temple built close to a pharaoh's tomb where religious ceremonies may be conducted in their honour.

Mummification The process of preserving a dead body as practised by the ancient Egyptians.

Mummy The body of a dead person or animal, carefully treated so that it will not decay.

Natron A salty mixture of sodium carbonate and sodium bicarbonate used in the mummification process.

Necropolis A large burial ground.

Nemes A striped headdress worn by some pharaohs.

New Kingdom Period of ancient Egyptian history roughly covering the years 1550 BC–1070 BC.

Nile The river on the banks of

which the ancient Egyptian civilization emerged.

Nilometer Device invented by the ancient Egyptians to measure the depth of the river Nile flood.

Nomarch A local governor, responsible for the government of a particular province (or nome).

Nubia The region that lay immediately to the south of Egypt.

Oasis A fertile area with its own water supply, surrounded by desert.

Obelisk A tall pointed stone pillar, usually decorated with hieroglyphics.

Old Kingdom Period of ancient Egyptian history roughly covering the years 2649 BC–2134 BC.

Palestine A large region on the southeast side of the Mediterranean Sea, with which the ancient Egyptians engaged in both trade and war.

Papyrus A marshland plant with tough fibres, from which the Egyptians made paper. Also used to describe a piece of paper (papyri in the plural).

Pectoral A large piece of jewellery worn on the chest by pharaohs and other rich people.

Peret The season in which ancient Egyptian farmers planted and grew their crops, roughly from November to February.

Pharaoh A king or ruler of ancient Egypt.

Predynastic period Period of ancient Egyptian history that predated the unification of the country.

Pylon A massive stone gatehouse marking the entrance to a temple, or dividing one part of a temple from the next.

Pyramid A massive triangular tomb with four sides, in which the body of a dead pharaoh may be buried.

Relief A carving of hieroglyphics or figures on stone.

Sarcophagus Stone coffin in which an inner wooden coffin containing a mummy would be placed.

Scarab Sacred symbol in the shape of a scarab beetle.

Scribe An Egyptian official who was able to read and write.

Scroll A lengthy roll of writing paper made of sheets of papyrus joined together.

Senet An ancient Egyptian board game resembling chess.

Shabti A model of a servant, which was buried with a dead man in the belief that it would come to life to serve his master in the afterlife.

Shaduf A simple lifting device used by farmers to raise water from the river Nile and transfer it to their fields.

Shemu The season in which ancient Egyptian farmers harvested their crops, roughly March to August.

Sistrum A type of rattle played by priestesses during religious ceremonies.

Sphinx A statue depicting a mythical creature with the body of a lion and the head of a man.

Stela A stone tablet bearing carvings or hieroglyphics or other figures.

Step pyramid A pyramid constructed in a series of stepped layers, each smaller than the one below and ending in a flat top.

Throwstick A hand-held hunting weapon resembling an aboriginal boomerang.

True pyramid A pyramid with smooth sides, ending in a point.

Underworld A mythical place to which the spirits of the dead travelled after death.

Upper Egypt The southern part of Egypt, including Thebes and the Valley of the Kings.

Vizier A senior scribe, second in power only to the pharaoh.

Wadjet A sacred symbol in the shape of the eye of Horus.

INDEX

Look out for further titles in the Collins Gem series.

Collins *gem*

Ancient Rome

From amphitheatres & aqueducts to gladiators & galley-slaves

Collins *gem*

Ancient Greece

From Drama and Democracy to Muses and Mythology

Collins *gem*

Pirates

From corsairs and cutlasses to parrots and planks

Collins *gem*

Royal Britain

Fascinating insight into British royal life